WATER USE AND R

Edited by David Newton and George Solt

INSTITUTION OF CHEMICAL ENGINEERS

Opinions expressed in this volume are those of the individual authors and not necessarily those of the Institution of Chemical Engineers.

All rights reserved. No part of this publication may be reproduced, stored in a retrieval system, or transmitted, in any form or by any means, electronic, mechanical, photocopying, recording or otherwise, without the prior permission of the copyright owner.

Published by
Institution of Chemical Engineers,
Davis Building,
165–189 Railway Terrace,
Rugby, Warwickshire CV21 3HQ, UK.

Copyright © 1994 Institution of Chemical Engineers
A Registered Charity

ISBN 0 85295 357 7

Cover material by courtesy of PowerGen.

Printed in the United Kingdom by Galliard (Printers) Ltd, Great Yarmouth.

PREFACE

Recent years have seen a rapid rise in water and effluent disposal costs. At the same time, concern has increased about the impact of industrial activity upon the environment.

The incentive to use water more efficiently and minimize effluent discharges has increased considerably. Costs of water, effluent treatment and off-site disposal, which have previously been regarded as insignificant or unavoidable are now, for many industries, becoming intolerably high and potentially prohibitive to commercial viability.

There is clearly considerable potential for the achievement of significant reductions in the water bills and effluent disposal costs of many companies. The range and effectiveness of technology applied has been developed rapidly in recent years, membrane technology being an outstanding example. This trend will certainly continue and probably accelerate, although whether the instinctively attractive concept of 'zero discharge' will ever be practical, let alone economical, remains to be seen.

A major part of the real problem lies in attitudes, often across all levels of the organization. There is still a wide perception that water is free, or at least very cheap. It is true that water is at least 100 times cheaper than most bulk chemicals, which is remarkable because it is also 100 times purer (non-potable water is even cheaper). On the other hand, the volume of water which is used in industry is prodigious, and raises a serious cost.

Another obstacle to water conservation within some companies is organizational. Often no individual holds clear accountability for the efficient use of water or the minimization of effluent discharges. Consequently, insufficient consideration is given to the efficient use of water when new processes are being designed. Plant operators are neither adequately trained in the efficient use of water nor provided with appropriate facilities to monitor and control water usage.

Given a sound approach to water and liquid effluent problems, however, appropriate technology is usually either already available, or readily developed, to conserve water and minimize environmental impact, while yielding a tolerable financial return.

This book deals with all of these matters, including the application of water-pinch technology on a multi-process site, the application of generic methodology for water management and the development and application of generic methodology for water conservation. Many practical examples of major reductions in water usage and effluent discharge volumes are given, including those resulting from the classic 'Project Catalyst' in the Mersey basin.

The chapters of this book are based on papers presented at a conference organized by the North Western Branch of the Institution of Chemical Engineers in co-operation with the Institution's Specialist Water Group, and held at UMIST, Manchester on 19 October 1994. It is a pleasure to acknowledge the contributions of Ian McConvey, Tony Thompson and Mike Adams in devising and developing the programme for that meeting. They join us in thanking the authors for so enthusiastically contributing their valuable expertise to this book.

We are confident that the book will be a source of inspiration to all of those who are in any way concerned with the conservation of our most valuable resource and with protection of the environment.

David Newton and George Solt

THE AUTHORS

1. Is zero aqueous discharge a practical option?
   Rob Terrell and Mike Holmes *(ICI Engineering, UK)*

2. Establishing a strategy for effective water source management and control
   Colin Appleyard and James Lander *(Eckenfelder & Binnie Ltd, UK)*

3. Defining a strategy for fresh water and wastewater minimization using water-pinch analysis
   Eric Petela *(Linnhoff March Ltd, UK)*, Robin Smith *(Department of Process Integration, UMIST, UK)* and Ya-Ping Wang *(Linnhoff March Ltd, UK)*

4. Project Catalyst — a waste minimization demonstration project
   Simon Clouston *(WS Atkins Consultants Ltd, UK)*

5. Ion-selective membranes in effluent treatment and prevention
   Jan Tholen *(TIM, The Netherlands)*

6. Reverse osmosis membranes in cooling towers
   Stuart Ord *(ICI Chemicals & Polymers Ltd, UK)*

CONTENTS

PREFACE		iii
THE AUTHORS		v
1.	IS ZERO AQUEOUS DISCHARGE A PRACTICAL OPTION?	1
2.	ESTABLISHING A STRATEGY FOR EFFECTIVE WATER SOURCE MANAGEMENT AND CONTROL	23
3.	DEFINING A STRATEGY FOR FRESH WATER AND WASTEWATER MINIMIZATION USING WATER-PINCH ANALYSIS	39
4.	PROJECT CATALYST — A WASTE MINIMIZATION DEMONSTRATION PROJECT	47
5.	ION-SELECTIVE MEMBRANES IN EFFLUENT TREATMENT AND PREVENTION	61
6.	REVERSE OSMOSIS MEMBRANES IN COOLING TOWERS	73

1. IS ZERO AQUEOUS DISCHARGE A PRACTICAL OPTION?
Rob Terrell and Mike Holmes

Saving water makes sense to everyone. Water is a natural resource that is becoming increasingly valued and there is a growing perception — not always supported by statistics — that the pattern of rainfall, and hence the supply of good quality water, is changing. At the same time, there is a growing awareness of the impact of pollution, an enhanced ability to measure minute concentrations of contaminants and a demand for ever-tightening water quality standards for both potable water and waters in the environment. The public is aware as never before and is demanding improvement.

'Zero discharge' seems likely to become the environmental target for the future on the basis that, if there is 'zero discharge', there can be no environmental impact and no pollution. The concept is instinctively supportable, but what is meant by 'zero discharge'? It has been variously interpreted as 'zero environmental impact discharge' and 'zero aqueous discharge' and there are concerns that, in an attempt to minimize aqueous waste, industry should not be forced into environmentally unacceptable practices, such as the production of large quantities of solid or gaseous waste, or into energy-intensive recovery processes.

This chapter examines whether zero aqueous discharge is achievable, practical, beneficial or economic. It also asks whether water management techniques to reduce water consumption form a more realistic target.

DEFINITION OF ZERO AQUEOUS DISCHARGE
For the purposes of this chapter, zero aqueous discharge means no aqueous discharge from any point on the manufacturing site. This means either the total recycling of water, including aqueous effluent, or the evaporation of any effluent to give a solid waste for landfill or other means of disposal. Domestic waste is excluded, although on sites with a biological treatment plant it may be possible to treat domestic waste successfully.

Water management is the name given to the integration and control of water systems so that both the usage and discharge of water is minimized. This includes all areas of water use — process and utilities such as cooling and water purification.

THE DRIVING FORCES
There are several driving forces to minimize water use which are common to all industries and geographical regions. The differences between the driving forces may be subtle, but the distinctions between them are very important to the industry involved and the attitude that industry displays towards reducing water consumption. It is common to find a combination of driving forces being used to support specific projects.

AVAILABILITY AND/OR QUALITY OF RAW WATER
Many regions of the world suffer from a chronic shortage of raw water of suitable quality for domestic or industrial use. Variations in rainfall patterns have reduced the availability of good quality surface waters, while ground waters may become increasing saline as more water is drawn from bore holes and wells. The availability of water for human consumption takes precedence over industrial use, so industry is forced to look for alternative raw water sources which are less pure, to recycle water — for example, from municipal effluent treatment plants — or to reduce its water consumption. This driving force is largely geographical rather than related to the industry. Examples include North America and Australia.

COST OF RAW WATER
Increasing demands for higher quality drinking water have led to very large capital investments by the suppliers of water to ensure removal of trace contaminants. These may be heavy metals, organic material and environmentally persistent chemicals such as pesticides, fungicides and insecticides. This increased investment has resulted in significantly higher water costs to consumers, both private and industrial, although the improvements in water quality are of little direct benefit to industrial users. This factor is of particular significance in Europe, where prices are expected to continue to rise in real terms into the next century. It is also significant in the developing countries where higher living standards increase the demand for better quality drinking water.

ABILITY TO DISCHARGE

Where surface and ground water have to be used for potable water, the discharge of industrial effluent into watercourses in the vicinity of abstraction points may be prohibited and 'zero aqueous discharge' may become a condition of planning permission. This is a largely geographical factor. It is of particular significance in countries with a large land mass, but may also be imposed on industrial companies who have special waste problems, in particular the need to dispose of toxic wastes which are persistent in the environment. North America has led the way in this area and Europe appears likely to follow.

DISCHARGE CONSENT LIMITS

Even if the discharge of aqueous effluent is permitted, this discharge is subject to increasingly stringent consent conditions imposed by the Regulatory Authorities. It is likely that discharge restrictions will become even tighter in the future, requiring more investment in effluent treatment plants if industry is to maintain its 'licence to operate'. Discharge limits are dependent on the quality and size of the receiving water; thus inland sites on small watercourses will suffer many more restrictions than sites on estuaries or by the sea. The imposition of tighter discharge consent limits follows industrialization, with North America, Europe and Australia leading the way. There is evidence that South East Asia will be moving to match these conditions.

CAPITAL COSTS OF EFFLUENT TREATMENT PLANT

Investment in new effluent treatment plant is viewed as non-productive but essential by most industrial sectors. But the increasing costs of effluent treatment provide an added incentive for the reduction of volume and contamination of wastewaters.

Effluent treatment is commonly based on suspended solids removal and biological treatment, for which the basic plant design comprises large settlers and holding tanks, often allowing several days' holding capacity for wastewater. For this reason, the cost of such plant is largely determined by the volumetric throughput. Plants are built larger than required, to allow for upsets and uncertainties about the ability to treat the waste adequately. If it were possible to reduce the total volume of effluent — for example, by reducing the demand for water — significant capital cost savings should result.

RISING ENVIRONMENTAL COSTS
Companies are finding that their 'environmental costs', including the treatment of aqueous effluent, are rising significantly faster than other costs. In practice, perceived costs seldom reflect the real costs of effluent treatment, since cost data and analysis are weak in this area. To maintain a competitive position, it is essential to find ways of reducing the rate of increase, while still complying with consent conditions.

COMPANY IMAGE
For some companies, maintaining a 'green' image is seen to be important. Pharmaceutical companies and specialty chemical producers are very sensitive to this issue, believing that once the company image is damaged, it takes a long time to repair it. Companies with high volume, low margin products put less value on this factor, meeting consent conditions only when required by legislation and being prepared to risk fines rather than spend money on improvements.

MARKETING STRATEGY
Some companies are prepared to use the environment as part of their marketing strategy. Instead of just protecting their image by meeting all consent conditions, these companies believe that there is a competitive advantage to be gained by being 'greener' than their competitors. They are prepared to invest to anticipate legislation, rather than react when forced. This factor is industry-dependent.

IS ZERO AQUEOUS DISCHARGE ACHIEVABLE?
Increasing numbers of examples of zero aqueous discharge are reported in a variety of publications. The majority of these are for new plants in the USA, where environmental legislation is particularly tight. A closer look at the examples reveals that they are largely restricted to a limited number of industries, to small factories or to individual plants.

The prime examples are cogeneration power plants in the US Mid West, where:
- water is in short supply;
- the degree of contamination, which might otherwise complicate the recycle process, is low;
- land for lagoons and landfill is cheap and readily available;
- evaporation is favoured over precipitation;

- there is a plentiful supply of low grade heat and power to operate evaporation processes.

Several examples are pulp and paper mills which have been active in reducing their water use — by as much as 90% over the past 20 years — and who are now actively seeking to 'close up' the mills in response to public concern about pollution of receiving waters. Historically such factories have had a large requirement for good quality river or lake waters, but there are now operating examples of zero aqueous discharge mills in Canada and New Zealand. Integration of utility water use with process water use remains one of the outstanding challenges.

Fewer examples of large chemical and petrochemical sites are reported. They tend to be large users of water with complex and highly integrated plants where contamination with organic pollutants is unavoidable.

The conclusion must be that zero aqueous discharge is technically achievable, at least for new plants in selected industries where the economic or operational driving forces are strong enough to justify the necessary investment.

IS ZERO AQUEOUS DISCHARGE PRACTICAL?
The practicality of operating with zero aqueous discharge depends on:
- the nature of the contamination;
- the operability of the unit operations involved;
- integration of the water streams;
- understanding of the water chemistry.

Several new plants have been designed with zero aqueous discharge as an aim, but it has not been possible to close the water systems because of poorer-than-design performance from the unit operations used for recovering and recycling water. This can be due to a number of causes, but is often related to a poor understanding of the impact of water chemistry on separation processes or the impact of the recycled stream on product quality or plant operation.

Examples of problems experienced include:
- extensive microbiological growth throughout the water system, resulting in blockages, product contamination, smell, discolouration, etc;
- scaling of filters and membranes or at points where streams from different sources are mixed;
- particulate fouling due to inadequate specification of pretreatment processes;
- corrosion of pipework, giving both materials failures and the presence of dissolved and particulate iron in the recovered water.

All of these problems can be predicted and avoided if there is an understanding of the water chemistry. However, the track record for control of cooling systems, boilers and water purification plant does not augur well for the future management of such highly integrated water systems.

IS ZERO AQUEOUS DISCHARGE BENEFICIAL?

There are some obvious benefits which derive from achieving zero aqueous discharge:
- reduced water costs — for extraction licence, pumping costs or potable water costs;
- freedom from discharge consent limits on both flow and contamination;
- maintaining a licence to operate — both in the immediate and the long-term future.

There are also some less obvious benefits which may be of significant value. An example is the ability to build new factories close to good distribution lines or markets, thereby reducing costs of logistics, rather than having to build them close to rivers, lakes or the sea. In addition, the ability to recycle water may make the quality of incoming raw water a less critical issue.

While there may be clear environmental benefits in the local area, the overall environmental benefit may be more difficult to determine. Which is better — to allow a small quantity of aqueous discharge, or to burn fossil fuels to generate the heat to evaporate an aqueous effluent stream to dryness? Is thegreenhouse effect more important than putting a small quantity of naturally-occurring minerals to drain? Is landfill the best way to dispose of our waste or are we building mountains of waste for future generations? Life cycle analysis can be used to give an indication of the relative merits of reducing water consumption to a practical minimum and of trying to achieve zero aqueous discharge, but this procedure is relatively new and the results are open to various interpretations. Even when this work has been done, local environmental pressure groups may be less interested in the greater environmental benefit than in getting rid of an unsightly discharge.

IS ZERO AQUEOUS DISCHARGE ECONOMIC?

Ultimately, any investment has to be justified on economic grounds:
- is there an acceptable financial return?

- does this affect the ability to maintain a licence to operate?
- will the company image or marketing position be affected so that operations become unprofitable?
- are there acceptable alternatives?

CRITERIA FOR INVESTMENT
Each of the identified driving forces has an economic basis which needs to be compared with the cost of end-of-pipe treatment. While no company would be prepared to spend money unless there were a tangible financial benefit, it is clear that companies may not know the real costs of water handling. The way in which costs associated with water use and effluent treatment are allocated vary quite widely, as does the required payback period for investment. As a result, even within the same industry, some companies have made significantly more progress in reducing water consumption than others.

The criteria for investment to meet imposed discharge consent conditions are quite different, since meeting the conditions is seen as an essential, if undesirable, cost in order to continue production. In these cases, no clear identification of the cost benefits of reducing water consumption exists. Indeed, projects entered into under environmental pressure are viewed as a burden, rather than an opportunity, with the result that only the minimum work is carried out and water reuse/recycle is not considered. This is partly the result of having insufficient time to evaluate minimization opportunities prior to the design of treatment facilities.

Where a commercial advantage in being 'green' is identified, projects are approached with a more positive and open attitude and investment and payback criteria viewed in a different way.

REDUCING WATER CONSUMPTION VERSUS END-OF-PIPE TREATMENT
For new plants, it is relatively easy to incorporate water management philosophy at the design stage, given that the technical uncertainties can be overcome. The incremental cost of modifying the process may be compared with the long-term operational cost savings and capital costs associated with an effluent treatment facility, and an assessment of the ability of the plant to meet future environmental legislation.

For existing plants, the potential benefits of reducing water consumption may be offset by the costs associated with the installation of storage tanks and long pipe runs. Modifications to plant pipework and supporting steelwork may be expensive and difficult to manage and there is a natural reluctance to interfere within existing plant battery units. In these circumstances, and in the absence of clear information about the direction and timing of future environmental legislation, the attractions of a stand-alone effluent treatment plant are evident. There may be considerable benefits to be drawn, however, from selective water management schemes, both in operational cost savings and, technically, in meeting discharge consent limits.

ENVIRONMENTAL LEGISLATION

As environmental legislation grows, an increasing similarity can be seen in the standards set in different countries, the pollutants identified and the approach taken. Common features in UK/EC, USA and even Japanese legislation include:
- objectives for water quality are usually based on the use of the receiving water;
- standards for the concentrations of a wide range of substances in the water are related to the objective of water quality;
- discharge consents or permits provide the practical detail required to achieve the required standards in the receiving water.

Future discharge restrictions imposed will depend on many factors over which industry may have little or no control — for example, political changes, lobby groups, environmental studies, the development of suitable test methods and changing timescales.

It is reasonable to assume, however, that environmental standards will continue to become more common across geographical boundaries as environmental studies are carried out and legislative bodies follow each other in the limits adopted. Even where legislation is not in place, pressure from lobby groups is threatening to impose similar standards across geographical boundaries by restricting trade unless the manufacturing processes conform to local environmental legislation.

USE OF WATER BY INDUSTRY

Different industries regard water, and hence water management, in different ways. At one extreme is the oil and refining industry, for which water is solely a utility and very little is used for process duties. At the other extreme is the pulp and paper industry, for which water is an integral part of the process, entering with the raw materials and, historically, being used in copious quantities at every stage; for this industry sector, water management refers only to process water use and excludes utility use. The chemical industry includes both extremes — water is both a utility and a process fluid — and the industry regards water management as the integration of the two duties.

USE OF WATER IN THE CHEMICAL INDUSTRY
When considering water use, the chemical industry is best regarded as a collection of dissimilar but related industries. The production units involved vary from the processing of heavy organic chemicals and oil derivatives through the production of basic inorganic chemicals such as alkalis and acids and the production of highly complex specialty chemicals to the electrolysis of sodium chloride brine to produce chlorine. Even within specific related areas, the processes may be very different. In the production of plastics, it is essential for some processes that water is excluded if the product is to have the desired properties; for other processes, polymerization is carried out in emulsions in water and the quality of the water is critical for product quality. Plants producing the same product may use different routes, depending upon the location of the plant, the availability of raw materials of the required quality, other production processes on the same site, transport and safety considerations, the age of the plant and environmental discharge permits.

Water can enter the process with the raw materials, it can be added or removed at different stages in the process and it can be added to the final product to suit specific end uses or customer requirements. It is not uncommon to purchase speciality chemical formulations which contain in excess of 90% water; acids and alkalis may be sold with 50 to 70% water content.

It is common to find several different production processes on a single site. This is due to the integrated nature of the processes and considerations of the cost and safety of transporting potentially hazardous materials between sites. Thus water use on any particular site varies widely — some plants may only use water as a utility, while other plants may use water throughout the process. For this reason, while there are general rules which may be followed, strategies for water management are unique to production sites. For the chemical industry, therefore, it is essential to concentrate on generic solutions. This means looking at the unit operations which may find application for a range of uses — the building blocks which can be used to achieve a water management scheme which is economic and, above all, robust in its operation.

TABLE 1.1
Uses of water in the chemical industry

Production processes	Utilities
In raw materials	Cooling systems
— brines	— once-through
— acids and alkalis	— open evaporative
Product washing	— closed systems
— filters	Steam raising
— centrifuges	— fossil fuel boilers
— final product	— process boilers
Product formulation	Water purification plant
Reaction medium	— process use
Solvent	— boiler use
	Condensate recovery
Cleaning and housekeeping	**Other duties**
Cleaning of batch vessels	Firefighting
General plant cleaning	

Chemical production sites have traditionally been located where there is a plentiful supply of cheap water and where the discharge of water has been relatively unrestricted. Environmental, safety and marketing pressures are now changing this picture and this, in turn, demands more consideration of the way water is used and discharged. Coupled with the rising costs of water and an increasing awareness of the impact of water on production and product quality, the opportunities offered by water management are drawing more interest.

The different ways in which water is used are described in more detail in Table 1.1.

PRESENT TRENDS IN WATER MANAGEMENT AND THE IMMEDIATE FUTURE

The chemical industry has made little progress in recent years in reducing its water consumption. The introduction of open evaporative cooling systems was the last major step forward, but there has been little change over the last 20 years for three reasons:

- water is seen as a cheap, readily available commodity in inexhaustible supply;

- there is a perception that it is acceptable, even desirable, to dilute effluent as much as possible to meet environmental discharge limits;
- the diversity of the chemical industry makes it difficult to provide a meaningful measurement system by which progress may be judged — for example, m^3 water per tonne of product.

Progress has been made during this time to improve the control and management of existing water systems, but this has been driven by waterside failures of heat exchangers and boilers. Many arguments have been put forward to explain the lack of progress in other areas:
- concern about the impact of water reuse and recycle on product quality;
- a lack of understanding of the contaminants present in the water;
- the practice of mixing effluent streams rather than separating them;
- the perceived absence of reliable technology for the removal of contaminants;
- the absence of an adequate payback for investment.

In reality, there has been no culture to reduce water consumption and existing plants were not designed with water conservation in mind. Furthermore, accounting systems are designed to determine the cost of producing chemicals and cannot easily be applied to measure the true cost of using water.

It is very unlikely that complete water management schemes will be introduced for existing chemical plants, although partial schemes may be attractive. Major schemes will be limited by the cost of installing long pipe runs, storage tanks and pumps, but good payback may be achieved if product or raw materials can be recovered for recycling or for sale. The need to build new effluent treatment facilities and further restrictions on the discharge of specific pollutants may justify expenditure to reduce water consumption, but the current culture is to install large end-of-pipe facilities which are capable, at least in the short term, of meeting discharge consents. Investment is, therefore, likely to be limited over the next three to five years.

But if existing chemical plant is to maintain its licence to operate in the next century, the case for investment in water management schemes will inevitably become more attractive.

FUTURE TRENDS FOR WATER MANAGEMENT

There is an increasing recognition in the chemical industry that the reduction of water consumption is a necessary component of good environmental practice. Thus, while the case for investing in existing chemical plants may be limited, it is becoming recognized that new plant designs should consider not just water as a utility but the use of water as part of the environmental assessment.

Chemical plants have traditionally been designed for the process streams alone, with utilities and services added separately. For new plants, early consideration of water reuse and recycle schemes could bring changes to plant layout to incorporate the essential pipework, tanks and pumps and to install local separation unit operations capable of recovering both water and product. Essential modifications to cooling systems to allow the use of lower grade water would need to be made at this time. Water systems could be designed to supply water to specific duties in series, rather than in parallel as for existing plants. Measurement and control systems to safeguard product quality could be designed and installed as part of the plant control system.

The impact on the capital cost of the plant due to these modifications has not been assessed, but if effluent treatment is considered the increase, if any, is likely to be limited to a few percentage points. Operating costs would be expected to be reduced compared with existing plants. Of more importance is the ability to meet environmental legislation and maintain a licence to operate — something which end-of-pipe treatment may be unable to deliver.

METHODOLOGY OF APPROACH

CHARACTERIZATION OF WATER STREAMS
Each zero aqueous discharge or water management application is unique. The solution adopted depends on the raw water quality, the production processes in operation, the potential for reuse and recycle, the layout of the site, the geographical location and the receiving water stream. But there are generic solutions which may be applied with reasonable confidence if the water quality and flows can be accurately characterized.

In all the successful applications of water management reported, the first and most crucial step was a survey of the various water streams throughout the factory. After all, what cannot be measured cannot be controlled. The survey has to identify both the nature and concentration of all the contaminants and the mineral content of the water. The latter information is rarely available since it is of no value if the water is to be discharged directly, but it is of critical importance if separation processes are to be used as part of a water management scheme. The survey also has to identify the water flows for both steady state and upset conditions, since any treatment process has to be capable of operation under all possible conditions.

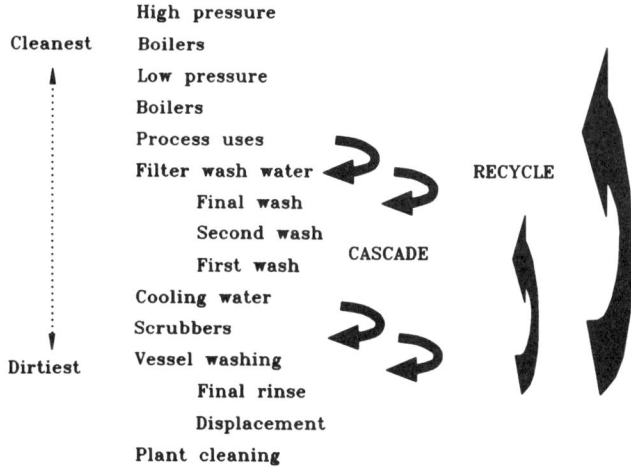

Figure 1.1   Classification of water duties.

The major water treatment companies and the suppliers of treatment chemicals for boilers, cooling systems and effluent treatment plants have all established techniques for this work and have developed computer models based on mass balances, so that an accurate picture of site operation may be put together. Such a model can be combined with simple functional and economic models of the treatment technologies which might be used to identify the most likely options. Then pilot plant studies determine the impact of contaminant flows and concentrations.

MATCHING WATER QUALITY TO DUTY

For most plant duties, the water used is of much higher quality than is required. Water for steam raising, or for specific process applications such as pharmaceutical or electronic component manufacture, is purified to required standards, but the majority of water is for duties where a much lower quality of water would be adequate. The reason for this 'parallel' use of water has been the cost of piping different grades of water throughout the plant. Alternative approaches based on matching water quality with process requirements are available (Figure 1.1).

If water management schemes are put in place, it becomes possible to use and reuse water of a quality appropriate to the duty. For this reason, the initial survey of the plant has to establish the water duties to ensure optimum water reuse without purification. Water management is not about collecting and treating the final effluent so that it can be returned to the front end of the plant to be used again. It is essential to look for ways in which water can be reused without further purification.

HIERARCHY OF ACTIVITIES

It is not possible to move from the present state to 'zero aqueous discharge' in one step. The complexities of most water systems mean that improvements have to be made progressively, starting with the cheapest and most beneficial. The various components of the water management scheme mentioned above have identifiable savings associated with them. For the chemical and refining industries, where both process and utility uses of water are included, the potential savings are as follows:

- reductions in uncontrolled use, for example poor housekeeping   20–30%
- improvements in control/management of existing water systems   20–30%
- reuse of water without treatment at other points in the process   10–20%
- treatment and recycle of water for further use in the process   10–20%
- improvements in process design so that water is not required   10–20%

Note that these estimates for potential water savings are only indicative, and depend on the particular industry and current practice. They are also dependent on the balance of water use for process and utility duties.

These actions represent a hierarchy of activities, from the simplest and cheapest to the most difficult and expensive. There is some overlap between the activities — that is, it will not necessarily be more expensive to reuse water than to improve the control of existing systems, but it will be difficult to get support for any investments to save water while there are seen to be excessive losses of water from existing systems. Figure 1.2 indicates the relationship between capital expenditure and potential water savings.

TECHNOLOGY FOR CHANGE

Most technologies applied in this field are non-destructive separation technologies. As such, they do not solve problems, but merely move them from one place to another. If a stream to be treated is contaminated with more than one material, then the separation process may only serve to create more problems. For this reason, the best place to install such unit operations is at the point of production of the waste material. This offers the further possibility of being able to recycle recovered material as well as the water.

The only destruction techniques available are biological treatment, oxidation and incineration, but these techniques also have disadvantages, such as the production of sludges, unwanted byproducts or the excessive use of energy. A water management process is likely to involve the use of both separation and destruction unit operations.

AVAILABILITY OF TECHNOLOGY

Water management is not technology limited. There is a wide range of separation technologies currently available, or in the process of development, which can be used for the partial or complete removal of contaminants and/or mineral salts from water so that it can be used again.

The most difficult area is the choice of technology for a particular duty. Many applications will require a combination of two or more unit operations, rather than just one. For most duties there will be a range of equipment which could be used; understanding the technical capabilities of the processes and the

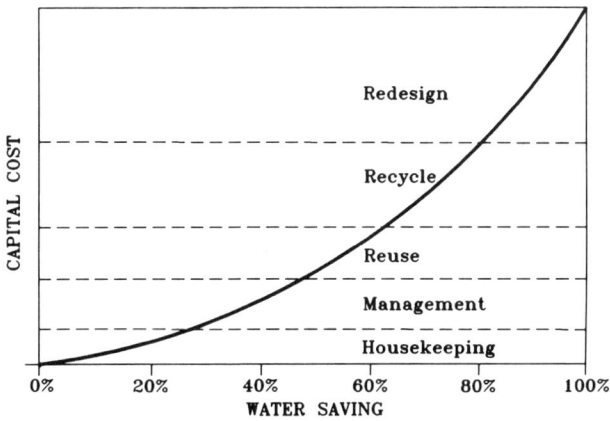

Figure 1.2   Cost of implementation.

TABLE 1.2
Technologies for suspended solids removal

Chemically-assisted	Mechanical	Membranes
Sedimentation	Centrifugal filters	Microfiltration
Coagulation	Belt filters	Ultrafiltration
Flocculation	Sand filters	Nanofiltration
Precipitation	Multimedia filters	Reverse osmosis
Softening	Cartridge filters	
Flotation	Fibre mat filters	

capital and operating costs of the equipment will be critical. For example, suspended solids removal may be achieved by a wide variety of techniques (Table 1.2).

One important consideration is the physical state of the removed components and whether they can be returned to the process. Most separation processes are designed to recover one phase rather than both. The removed material may not be in a phase which is readily recoverable, due to a change in phase, contamination with other materials or dilution. The application of non-intrusive separation technologies such as membranes may allow cheap and easy recovery of both the water and the removed phase, thus increasing the economic viability of the process.

Simple examples of this process would be:
- the recovery of paints from vessel washings;
- the recovery of dyestuffs from filter washings;
- the recovery of organic components after pervaporation.

Traditionally, water treatment and — in particular — effluent treatment have relied on large settlers and holding tanks to even out flows and concentrations and allow time for solids to settle. Water management requires a move to rely more on process engineering than on civil engineering. It is necessary to apply solutions that can be integrated into the process, rather than put after the process. Above all, it requires the use of good control technology.

IDENTIFICATION OF COMMON PROBLEMS

The available technologies can, with development, be used to treat a variety of problems. But the economic case for using one particular technology to address a range of problems may vary significantly. It is important to classify the problems and identify the potential solutions in order to make valid comparisons. Table 1.3 indicates which technologies might be used to address specific problems.

It is clear from the industries studied that, although there are problems which are specific to each industry, there are also many problems which are

TABLE 1.3
Technologies for water management

Suspended solids, oils	**Ammonia**
Gravity separation	Steam stripping
Coagulation	Nitrification
Flocculation, separation	Ion exchange
Flotation	Air stripping
Membranes	Breakpoint chlorination
Filtration	
Phenols	**Dissolved organics**
Solvent extraction	Aerobic treatment
Wet air oxidation	Anaerobic treatment
Biological oxidation	Chemical oxidation
Activated carbon	Activated carbon
Chemical oxidation	Wet air oxidation
	Incineration
Cyanide, thiocyanate	**Heavy metals**
Steam stripping	Chemical precipitation
Biological oxidation	Ion exchange
Alkaline chlorination	Activated alumina
Ion exchange	Reverse osmosis
Chemical precipitation	Electrodialysis
Dissolved solids	**Sludge**
Evaporation	Filtration
Reverse osmosis	Centrifuge
Electrodialysis	Incineration
Crystallization	Drying
Spray drying	Stabilization
	Landfill

TABLE 1.4
Performance of membranes

Technology	Particulates	Colloids	Pyrogens/ viruses	Organic molecules	Dissolved ions
Microfiltration	>0.1 μm	No	No	No	No
Ultrafiltration	>0.002 μm	Yes	Some	>2000 MW	No
Nanofiltration	>0.0005 μm	Yes	Yes	>100 MW	Divalent
Reverse osmosis	>0.0001 μm	Yes	Yes	>30 MW	Monovalent

common to all — for example, the removal of suspended solids, dissolved solids and heavy metals. It is therefore possible to learn from the experiences of other industries.

FUTURE TECHNOLOGY NEEDS
At the present stage of development of water management, it is difficult to identify what new technologies will be required, other than in the field of control and measurement. In the short term, effort will be concentrated on developing existing technology to suit new applications, rather than developing new technology. In the longer term, more problems may be identified which require radically new technology.

Ideally any new technology should be capable of separating the contaminants from water without requiring the addition of chemicals and without generating another waste stream. The production of a solid sludge as part of the separation process — for example, from clarification, heavy metals removal or biological treatment — simply moves the problem from one medium to another. Although it may be possible to reuse the water, the solid waste may become more difficult or expensive to dispose of. The addition of chemicals may contaminate either the water or the component being removed, making reuse more difficult due to the risk of contaminating the product.

Water management will focus on clean technologies — for example, the use of membranes to remove particulates, colloidal material, organic molecules and even dissolved salts with little or no addition of further chemicals (Table 1.4).

While membranes are not applicable in all cases, it is difficult to consider a water management scheme in which membranes do not figure. However, the performance of currently available membranes is dependent on the nature of the solids being removed and the presence of organic solvents, oxidizing agents (for example, chlorine), metals (for example, iron or manganese), pH, sparingly soluble ions (for example, strontium or barium) and silica. Pretreatment of the stream may be necessary if the membrane is to give an acceptable performance without fouling — for example, a combination of microfiltration and reverse osmosis. If membranes are to find wider application, the development of new, fouling- and chlorine-resistant membranes is a priority.

INSTRUMENTATION AND MEASUREMENT

One major concern is the impact of the recovered water stream on the process. This is particularly important when water is being transferred between unit operations within a process or between utilities, such as cooling water, and the process. There is little value in water reuse if the product quality cannot be maintained or if it becomes contaminated or discoloured, if scaling of plant items occurs or if the materials of construction start to corrode. These are problems which can be addressed if the water quality requirements have been identified and the correct technology used.

Plant operation is rarely steady for long periods, so it is essential that any recycle streams are continuously monitored for quality to avoid problems. Instruments are available for the more basic measurements required — for pH, conductivity and specific ions. In many cases there will be a requirement to measure low concentrations of organic molecules in the recycle streams. There are instruments available which are capable of performing this function, but there will need to be further developments in sensor technology if low cost, reliable and maintenance-free instruments are to become available.

CONCLUSIONS

In an ideal world, processing industries would not produce any waste material. Given that this is unachievable, at least for existing processes and technology, the next best aim must be to prevent the discharge of 'harmful' waste material to the environment — although the definition of what is 'harmful' is a topic which might occupy a lifetime.

The worthy aim of 'zero aqueous discharge' can only be achieved at the expense of increasing discharges to land or to the air or by the excessive use of energy for destruction of waste materials — all of questionable environmental benefit. It may be possible for some industries to achieve this goal relatively easily — for example, where the concentration and nature of the pollutants make them easy to recover, recycle or destroy, or where there is an excess of cheap waste energy available for destruction. The diversity of the chemical industry — the nature of the chemicals, the integration of plants on large sites and the lack of investment in new production facilities — makes this a distant goal for most factories.

The application of water management techniques to reduce discharges and water consumption is, at worst, a step in the right direction. At best it offers an approach which may lead us to a 'zero environmental impact discharge'. The key to reducing aqueous waste lies in managing the use of water through the production process to ensure that water is not used unnecessarily, that it is reused where appropriate and recycled after treatment where possible. By this method, it is possible to ensure that the minimum possible water passes forward to the effluent treatment plant for final purification and discharge. Effluent reduction through water management should be, in the longer term, inherently cheaper and more attractive than end-of-pipe effluent treatment.

This chapter indicates where reductions in water savings may be made and offers an approach and methodology by which opportunities may be identified and categorized. Water management is not technology-limited, but may be limited by experience or application of the technology for water systems. One significant area for development is in the field of measurement where more reliable sensors are required if concerns about the impact on product quality are to be overcome. However, do not be misled into believing that water management is easy. Like any chemical plant it needs to be carefully specified and operated and take full account of the water chemistry if it is to work successfully.

REFERENCES IN CHAPTER 1
No specific references are identified in the text, but the following were used as part of a study into water management carried out during 1993.

CONCAWE, Trends in oil discharged with aqueous effluent from refineries in western Europe — 1990 survey, *Report no. 1/92*.

Foster Wheeler Energy Limited, Private internal communication.

Jantunen, E., Lindholm, G., Lindroos, M., Paavola, A., Parkkonen, U., Pusa, R. and Söderström, M., 1992, The effluent-free newsprint mill, *Paperi ja Puu (Paper and Timber)*, 74 (1): 41.

Knorr, P. and Fromson, D., 1993, World-scale production of BCTMP with zero liquid effluent, *18th International Mechanical Pulping Conference, Oslo, June 15–17 1993*, 428–432.

Myréen, B., 1993, Recirculation of bleach plant waste liquors, *Tappi Environmental Conference*, 609–615.

Wang, Y.P. and Smith, R., 1993, Wastewater minimization, submitted to *Chem Eng Sci*.

The authors are indebted to many private communications held with numerous companies who are interested in this topic and have been prepared to share their views. These companies include Amoco, Betz Ltd, DuPont Nemours Inc, Grace Dearborn Ltd, Mobil, Nalco Ltd, Shell, Sun Oil and Zeneca, plus several Finnish pulp and paper companies and the Finnish Pulp and Paper Reseach Institute.

2. ESTABLISHING A STRATEGY FOR EFFECTIVE WATER SOURCE MANAGEMENT AND CONTROL

Colin Appleyard and James Lander

This chapter defines the means by which efficient use and reuse of water can be achieved in industrial applications. In this context the critical link between source management and control of cost-effective end-of-pipe effluent treatment is discussed. Both aspects involve the implementation of technical, management and operational initiatives to identify opportunities for beneficial use and reuse of water, materials recovery and general process optimization. They also point to reduction in end-of-pipe treatment requirements. A generic methodology for implementation of efficient water usage is outlined, covering aspects of management commitment and structuring, auditing, evaluation, implementation, training and monitoring. Case studies illustrate the concepts and techniques employed.

BACKGROUND

Most industrial wastewaters require some form of treatment prior to discharge to sewer or watercourse. The treatment can be extensive, involving a range of unit processes (physical, chemical and biochemical).

For many years much industrial pollution control has been 'end-of-pipe' and a wide range of unit processes have been developed to service the needs of industry.

Such end-of-pipe systems range from low to high intensity, from low to high technology and from low to high cost. Most are destructive processes in that they provide no return to the operating company in terms of increased product yield or lower operating cost, except in those circumstances where reduced charges then apply for discharge to municipal sewer.

In all cases the size (and hence the cost) of end-of-pipe treatment bears a direct relationship first to the volume of wastewater to be treated and second to the concentration of pollutants contained in the discharge. For example, the size of most physical-chemical reactors (balancing, neutralization, flocculation, sedimentation, flotation, oxidation, reduction, etc) is determined by hydraulic factors such as surface loading rate and/or retention time.

On the other hand, the size of most biochemical reactors is determined by pollution load — for example, kg chemical oxygen demand (COD) per kg of mixed liquor volatile suspended solids (MLVSS) per day in the case of suspended growth type systems, and kg COD per m^3 of media or reactor volume in the case of fixed film type systems.

Reduction of emissions by action at source can, therefore, have a significant impact on the size and hence cost of an end-of-pipe treatment system. On this basis it should be established practice in industry that no capital expenditure for end-of-pipe treatment should be made until all waste reduction opportunities have been exhausted. This has not often been the case, and many treatment plants have been built which are both larger and more complicated than necessary.

Accordingly, for direct cost benefit reasons, much more emphasis has to be placed on source management and control. It is a necessary first step to reduce the extent of end-of-pipe treatment to a minimum.

The enactment of the UK Environmental Protection Act in 1990 has established the requirement to justify the integrity of the very manufacturing operations themselves, as outlined by the HMIP Guideline notes to Chief Inspectors for integrated pollution control and the use of inherent cleaner technology. For many industries selection of end-of-pipe treatment alone is insufficient to meet the regulatory requirements. In such cases all new processes, and substantial changes to existing processes, must be assessed for authorization to operate. Further consolidation of this concept is expected from the development of European Union integrated pollution control directives.

Source management and control can be defined as 'the development of a full understanding of the nature of all waste streams (aqueous, gaseous or solid) and the exact circumstances by which they are generated in order to eliminate or minimize pollution before it arises'.

THE BENEFITS OF SOURCE MANAGEMENT AND CONTROL

The essential components of source management and control embrace a number of key technical, management and operational initiatives (see Figure 2.1).

Key technical initiatives involve identification of opportunities for:
- application of 'cleaner' processes or processing methods;
- enhanced housekeeping practices;
- water conservation, including reuse and recycle;

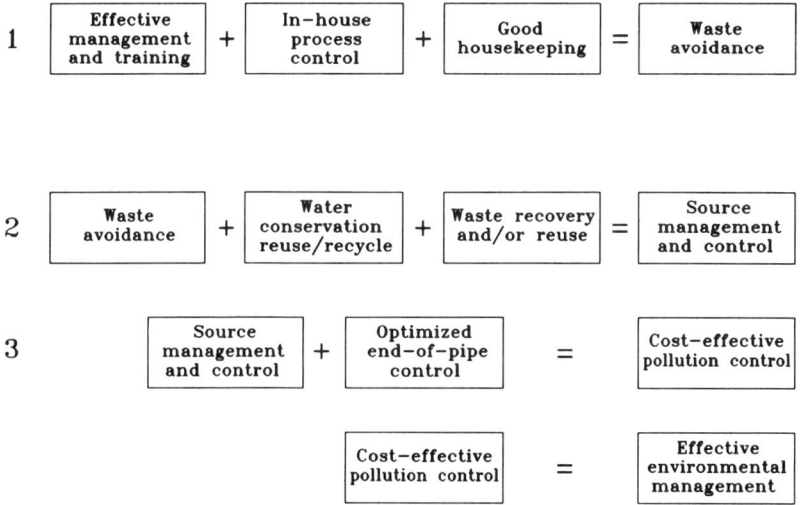

Figure 2.1 Industrial water and wastewater management.

- waste avoidance or minimization;
- materials recovery and/or reuse.

The potential benefits to be obtained in achieving optimum source management and control include:
- reduced water abstraction and treatment prior to use;
- enhanced product yield and productivity;
- reduced raw materials inventory;
- reduced end-of-pipe treatment requirement;
- improved regulatory compliance;
- reduction of risks (product loss, hazards, health and safety, contamination);
- potential for manpower savings and reduced land requirements;
- competitive advantages and improved market position;
- improved public relations/perception;
- improved status of stakeholders.

Potential benefits must be identified industry by industry — such potential benefits include both direct process benefits (primary benefits) and basic infrastructure benefits (secondary benefits).

In support of the concept of source management and control, experience shows that 30 to 60% of pollution can often be eliminated by reduction of waste and emissions at source.

Further support for source management and control include the arguments that:
- industry can maximize profits by increasing efficiency while at the same time maintaining environmental concerns. More often than not, external pressures from competitors is a principal motivating factor;
- pollution prevention is more environmentally effective, more technically sound and more economic than conventional controls;
- often little or no capital expenditure is required — just tightening up of procedures can produce major financial and environmental benefits. The further that measures for water management are developed and enhanced, however, so the cost to implement inevitably rises.

Technological modifications to existing plant and the installation of new production lines often provide the greatest opportunities, especially the latter where environmental criteria can be built into the design. While capital investment may be required, the cost is directly related to the improvement of manufacturing productivity and so payback should be attractive. Modification of the product itself should also be considered.

Source management and control in isolation is not always sufficient to achieve the overall objective of cost-effective pollution control and environmental management. What is required is detailed consideration of optimized end-of-pipe control for the irreducible minimum of wastes, in addition to source management and control.

A GENERIC METHODOLOGY FOR SOURCE MANAGEMENT AND CONTROL

The application of a generic methodology must be construed purely as a loose framework for the implementation of measures to determine the most effective use and reuse of water and other resources. A number of optimizing packages and other information management systems may be available for use, but such resources are only 'tools-of-the-trade'.

The key to the success of any investigation is a fundamental knowledge of the process technology and the very specific site circumstances. Consequently, an external party such as a consultant may provide an objective,

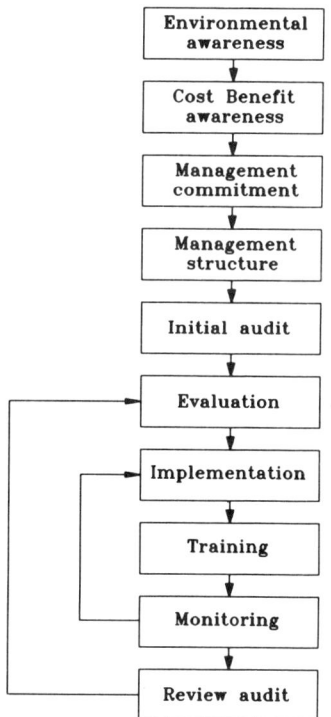

Figure 2.2 Procedures to achieve cost savings through source management and control.

impartial and enquiring approach (in the sense of hands not being tied) with specialist expertise and secure process knowledge. The committed involvement of, and transfer of knowledge to, the manufacturer will also be crucial; the manufacturer has an intimate understanding of the site operations and day-to-day anomalies.

The method of approach is outlined in brief in what follows and illustrated by an algorithm in Figure 2.2. The approach bears close resemblance to the initiation and development of an environmental management system (as exemplified by BS 7750) in following a 'control loop' pattern. Source management and control measures, in line with other loss prevention activities, can certainly play a key role in the development of a Total Quality Management system.

MANAGEMENT COMMITMENT

Commitment comes from the most senior level in a company and must be effectively embraced by all levels. To ensure the success of waste minimization and other environmental management activities, it is essential that the management commit their company to achieving effective pollution control by setting out environmental objectives, probably as part of a 'mission statement' and/or quality management system. Performance targets for water, materials and pollution emissions are set out. 'Procedures' are prepared establishing the management structure, operational procedures, training programmes and the schedule and methodology for audits to review compliance to the environmental objectives and performance targets. But such activities should be subject to reviews and development.

MANAGEMENT STRUCTURE

Often it is the maintenance manager who is given the additional responsibility for environmental protection, with the environmental manager reporting to him. More effective source management and control can be achieved by the environmental manager reporting directly to the production manager who is also responsible for the activities of the plant managers. In this way improved management interaction between production operations and pollution abatement is possible. By contrast the traditional alternative can be a somewhat tortuous management path.

INITIAL AUDIT

An audit of the site provides one of the most valuable tools in the evaluation of the optimum solution for site-specific source management and control. Audits are best carried out by, or in association with, third parties such as consultants experienced in such techniques. A joint initiative involving both consultant's and client's process engineers is the most effective approach. The audit is used to assess risks and identify opportunities for cost savings, etc. (The costs incurred by the audit are normally rapidly offset by the savings achieved.)

Audits, by their very nature, tend to be inquisitive and intrusive. Consequently the skill of auditors is to encourage the cooperation of operations staff. Staff training to explain the objectives, targets and methods for such investigations should be carried out at an early stage. Prior to the audit significant effort must be made to identify processes and collect data from the plant by undertaking a survey of the site. On large sites, the audit may only focus on certain areas of concern — a complete, exhaustive audit may not be practicable or necessary in many cases. Once this preparation has been completed and process flow diagrams drawn up, mass balances for the process(es) need to be produced. Raw materials, water usage, utilities, process outputs and discharges are determined and throughputs recorded. Input and output data are assembled and the mass balance is constructed and refined. Existing practices and procedures including methods of storage, handling and maintenance should also be recorded.

With this data base completed, a number of loss prevention methodologies can be employed (such as hazard and operability studies and hazard analysis) to target problem areas and assess their relative risks. Opportunities for cost savings are identified.

In terms of water usage on site, the audit considers the condition of water entering the system, its utilization within the plant and the nature of the effluent leaving the facility. Examples of such considerations are outlined here:

Incoming water
- What quantity and quality of feed water is required for each unit operation?
- What sources of water are available on site? Are they of suitable quantity and quality?
- What pretreatment may be required? Is the existing pretreatment system operating at its best? Can savings be made by using alternative equipment or operating procedures?
- Has the demand pattern changed recently or is it likely to do so in the future?

Production operations
- Can excessive demand peaks or washes be avoided or reduced — for example, to improve consistency of water treatment plant or to achieve water conservation in general. Can this be achieved by minimizing the rate at which water is used?
- Identify the pattern of water use for each separate unit process. Is the water over-treated prior to use? Can lower grade water be used satisfactorily?
- Identify the pattern of waste generation for each separate unit process. Why is it generated? Can entrainment of wastes into aqueous process streams be reduced or eliminated?

Wastewater
- Is it possible to segregate waste streams at source? Is it possible to recover materials from segregated streams using novel processes such as use of membranes?
- How does the cost of increasing the degree of treatment compare with charges arising from effluent discharge to sewer or watercourse?
- Would certain combinations of waste streams or individual waste streams benefit from pretreatment prior to combination with other streams for end-of-pipe treatment?

EVALUATION

The opportunities identified in the audit are then evaluated against a set of technical and environmental criteria which are weighted according to their order of importance and classified as either 'essential' or 'important' in a Kepner-Tregoe type approach[1]. The choice and relative importance of criteria can only be determined by the specific and unique circumstances of a particular plant or site. 'Essential' criteria might typically include meeting environmental objectives, productivity performance targets and corporate policy. Other factors such as technical risk, previous track record and ease of implementation/operation are also considered. Economic analysis can be carried out on each option in terms of capital and operating costs and the payback period. Each option is then ranked against each criteria in order of preference. The preferred option(s) can be determined by multiplying weighting times ranking and summing the totals. (A similar approach for selection of water pre-treatment plant or end-of-pipe treatment plant can also be adopted.)

IMPLEMENTATION

The preferred options are set against the environmental objectives and performance targets introduced by the management (see the section entitled 'Management Commitment'). Changes to operations, alterations to existing plant and installation of new plant (if necessary) are prioritized and then implemented. On commissioning these new operations, the plant data base is updated and documented.

TRAINING

Training needs reflect a commitment throughout the company and vary in nature according to the levels of responsibility held. Managers and supervisors require regular briefing on developments in environmental practices, policies and regulatory requirements. They should also become familiar with audit programmes as they will be participants in regular review audits.

Technical staff require training in the various techniques of source management and control, process optimization and end-of-pipe design optimization. They will be directly responsible for implementing the recommendations arising from audits. Operatives require an understanding of process optimization, good housekeeping practices, emergency procedures and emission control requirements.

The development of effective source management and control is a continuous process. Cooperation with other companies in the same and other areas of business, as well as with specialist consultants, water companies, regulatory bodies and the general public, can provide significant benefits in terms of technology transfer and dissemination of other companies' experience. It introduces a healthy cross-fertilization of ideas as well as access to specialist knowledge and services.

All parties should be made to realize the importance of regular objective auditing and their full cooperation in this activity, to be construed as an integral part of operations.

MONITORING

Monitoring is a vital component of the management of process performance and the control of water utilization. It may be discovered that existing records are inadequate to identify the true benefits open to a company. Consequently setting up an appropriate monitoring programme is a prerequisite to success in water and waste management. With modifications to the plant set in operation, routine checks of emission loads and concentrations and other crucial process streams should be programmed and recorded. It is likely that the new regime of operation will affect the maintenance requirements and evidence of this needs to be monitored closely.

REVIEW AUDIT

Resourcing must be made available for a regular programme of audits to be implemented at appropriate intervals in order to confirm adherence to the company's environmental objectives and performance targets as well as continued regulatory compliance. As in previous cases, these audits should be carried out by disinterested third parties or other internal departments not directly involved in the activities. Such audits also confirm the effectiveness of the management system. (However, here there is a danger of assessing the effectiveness of the management system rather than operation performance itself.) Results should be fully documented to allow subsequent inspection as, from time to time, third party audits may be required by customers or regulatory bodies.

The audit also provides opportunity for regular re-evaluation of processing methods and procedures, and may result in further changes as appropriate.

EXAMPLES OF SOURCE MANAGEMENT AND CONTROL IMPLEMENTATION

Savings to be made by successful implementation of source management and control vary according to the particular industry. Table 2.1 shows typical figures for reduction in pollutant loads as a result of source management and control measures. The potential for savings is also very site-specific. For example, modifications to segregate waste streams may be difficult or impracticable for older or more complex sites. Nevertheless, savings may be possible elsewhere,

TABLE 2.1
Typical case studies of successful source management and control

Industries	Before	After
Chemical:		
• volume (m^3/d)	5000	2700
• COD (t/d)	21	13
Hides and skins:		
• volume (m^3/d)	2600	1800
• BOD (t/d)	3.6	2.6
• TDS (t/d)	20	10
• SS (t/d)	4.83	3.7
Metal preparation and finishing:		
• volume (m^3/d)	450	270
• chromium (kg/d)	50	5
• TTM (kg/d)	180	85

COD = chemical oxygen demand; BOD = biological oxygen demand;
TDS = total dissolved solids; SS = suspended solids; TTM = total toxic metals.

TABLE 2.2
Optimization of water usage in batch and continuous chemical processes

Process type	Batch	Continuous
Process control	• Optimize mass balance • Input raw materials accurately • Control reactor conditions • Optimize phase separation • Slow down procedures • Reprocess off-specification material • Enhance reaction rates	• Control water to specific requirement • Side stream contaminant removal • Side stream process solution enhancement • Minimize 'drag out' factors • Replenish to prolong run times
Plant cleaning	• Optimize campaign working • Recover first flush • Reuse final rinse • Use clean-in-place techniques	• Reduce frequency • Use pressure techniques • Include automatic shut-off
Support utilities	• Recover spillage/detect leakages • Optimize cooling water management and stream usage	• Recycle cooling water • Reuse cooling water • Optimize stream usage • Consider energy recovery

such as in modifications to process control, plant cleaning and support utilities. Table 2.2 presents examples of actions that could be implemented for optimizing water usage in the case of batch and continuous chemical processes. The case history which follows presents examples.

Such measures provide opportunities for:
- water conservation/reuse/recycle;
- minimization or avoidance of waste;
- recovery or reuse of materials (metal salts, drag out, first flushes, off-specification product, energy, solvents);
- using alternative technologies or procedures to their best effect.

The economic benefits of alternative technologies in the context of a source management and control strategy are often very attractive in terms of savings and payback on investment. With payback periods often of less than one year, such opportunities cannot be ignored.

A CASE HISTORY
This case history presents the findings of a water conservation study at a military munitions research establishment. The primary goal of the study was to identify ways of reducing wastewater discharges substantially in preparation for the establishment's link to the local sewerage system. A target figure of 380 m^3/d was required compared to a typical average discharge of about 637 m^3/d.

The investigations consisted of three broad tasks. First, the existing water intake and discharge patterns at the centre were defined. Second, potentially feasible alternatives for reducing wastewater discharges were identified. Third, the potential alternatives were evaluated to select the most cost-effective means of reducing the centre's wastewater discharges.

EXISTING WATER INTAKE AND DISCHARGES
Nearly 60 significant water uses, other than for sanitary purposes, were identified during four on-site investigations involving a joint consultant/client team. The water distribution system was segregated into two distinct components, with abstraction from two points, sources A and B.

After use water was discharged in the following ways:
- to the facility's wastewater treatment system;
- to a local creek, through industrial waste outfalls;
- to the atmosphere;
- to the ground, both directly and indirectly.

A summary of the average water discharges is given in Table 2.3.

TABLE 2.3
Case history — summary of average water discharges

	m^3/d
Waste treatment	350
Industrial waste outfalls	280
Atmosphere	285
Ground, direct	48
Ground, indirect	7
Total	970

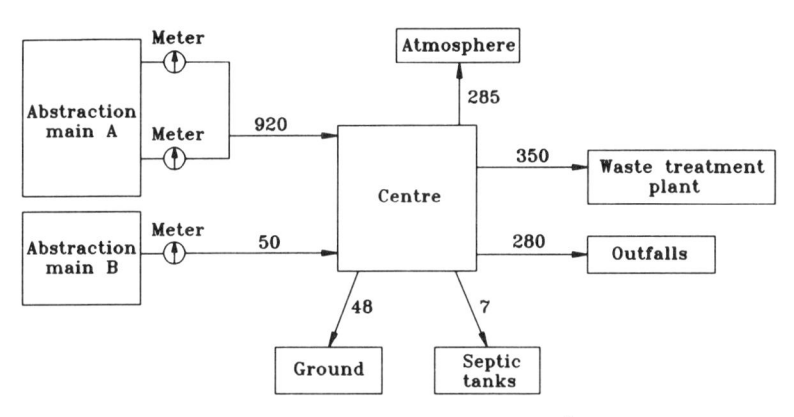

Figure 2.3 Water balance schematic (flows in units of m^3/d).

A water balance schematic is presented in Figure 2.3. Taking into consideration the evaporative losses from the cooling towers and water mains seepages and other losses, the total flow to sewer would be about 637 m^3/d. Overall water use was lower by up to about 25% during winter months due to reduced evaporative losses, watering of grounds and air conditioning load, but with little change in wastewater for discharge to sewer.

POTENTIALLY FEASIBLE WATER CONSERVATION ALTERNATIVES
The research-oriented industrial activities at the establishment created water use patterns which greatly reduced the feasibility (technical and economic) of many common water conservation methods.

The major areas where water conservation methods could potentially be technically and economically implemented included the following:
- reduction in use of once-through cooling water;
- selective reuse of certain cooling waters for process duties;
- cooling tower blowdown control;
- reduction of plating shop water use;
- water-saving sanitary fixtures;
- increased employee water conservation awareness.

ECONOMIC EVALUATION OF WATER CONSERVATION ALTERNATIVES
Cost estimates were prepared for each technically feasible water conservation alternative. To provide a common unit of comparison for all alternatives, the cost estimates for each alternative were divided by the projected water savings to determine the capital cost per unit water savings.

The initial economic evaluation was directed toward the goal of reducing projected wastewater discharges to sewer to 380 m^3/d or less. The economic evaluation indicated that the projected wastewater discharge to sewer could be reduced from approximately 637 m^3/d to approximately 356 m^3/d with a favourable savings/investment ratio (SIR) of 2.86 for this water conservation expenditure.

The SIRs of the remaining water conservation measures, beyond those necessary to achieve the 380 m^3/d wastewater goal, were also calculated. Two additional measures had SIRs greater than 1.0. Implementation of these two additional measures would reduce wastewater sewer discharges to approximately 327 m^3/d. Overall water savings of 310 m^3/d were identified. Consequently overall water use could be reduced from 970 m^3/d to a projected 660 m^3/d.

CASE HISTORY CONCLUSIONS
The establishment was able to reduce its projected wastewater discharge to sewer to less than 380 m^3/d. The water conservation measures associated with this wastewater reduction are summarized in Table 2.4.

The reduction in wastewater discharges enabled the establishment to discharge directly to sewer and, therefore, to abandon its existing wastewater treatment facilities and industrial waste outfalls direct to water course. Operating cost savings were realized, with abstraction substantially reduced due to the water conservation measures.

In addition to the water conservation measures listed in Table 2.4, the following measures were also pursued:
- increasing employee awareness of water conservation;
- providing a makeup water softening system as part of the upgrading of the other cooling tower system;
- identification and repair of leakages on the water supply pipelines around the site.

TABLE 2.4
Summary of recommended water conservation measures

Recommended water conservation measures	Projected water savings (m^3/d)
Reuse compressor cooling water	12
Repair leaking valve in Plastics Shop	5
Convert from once-through cooling water to recirculated supply from existing cooling towers.	15
Install a new cooling tower to serve equipment cooling needs in various buildings	140
Install thermostatically-controlled solenoid valve on once-through cooling water to the TCE degreaser in the Plating Shop	8
Install makeup water softening for cooling towers	60
Install conductivity controllers on cold water rinse tanks in Plating Shop, plus selected reuse opportunities	70
Total	310

CONCLUSION

Efficient use and reuse of water can be achieved through establishment of effective source management and control of water intake and raw materials as well as monitoring manufacturing process efficiency. This results in both conservation of water and minimization of waste, with substantial benefits in relation to direct cost savings and reduction in size (and hence cost) of end-of-pipe treatment systems. The success of such investigations is dependent on commitment throughout the organization as well as a sound knowledge of process technology and the unique circumstances of the particular site, to which any waste minimization measures must be tailored. The cost benefits and other benefits are often very substantial with low payback on investment. The greatest opportunities, however, are to be found when investing in new process plant and equipment, where efficient water use and environmental protection criteria can be built into the original design.

REFERENCES IN CHAPTER 2
1. Kepner, C.H. and Tregoe, B.B., 1981, *The New Rational Manager* (John Martin, London, UK). The authors developed methodologies for the effective management of information, personnel and other resources in problem solving, decision making and strategy planning.

3. DEFINING A STRATEGY FOR FRESH WATER AND WASTEWATER MINIMIZATION USING WATER-PINCH ANALYSIS

Eric Petela, Robin Smith and Ya-Ping Wang

To date, campaigns to reduce water use and wastewater production have generally used a 'cherry picking' approach. Whilst such techniques have been useful in identifying savings and improvements, there is usually some residual doubt that perhaps not all opportunities have been identified or that those identified may not have been the best opportunities.

Over the last two years UMIST and Linnhoff March have developed a new and systematic approach to identifying all water and wastewater reduction opportunities. The approach, termed 'water-pinch analysis', allows the setting of targets on a total-site basis for minimum water use and wastewater production. The technique can address any number of contaminants and quantifies the benefits for reuse, regeneration and recycling.

BACKGROUND TO THE APPROACH

On any industrial site, water is used for many and varied purposes. In the process industries water is used as boiler feed water, for cooling tower makeup and for various process requirements such as product washing, equipment washing and quenching. Whilst it is often possible to identify saving opportunities on individual water users, the engineer generally has little knowledge of how such savings benefit the site as a whole, or if indeed the savings identified are the best options in terms of cost and investment. What is needed is a systematic approach which allows the engineer to view the total site operation and to put the many potential water saving projects in the context of benefits achieved at the site boundary.

SINGLE WATER USER

In developing the approach to a total system analysis, consider a situation where water is being used to wash soluble impurities from a product[1]. The product enters the washing operation with an initial contamination level of $C_{s\,in}$ and must be cleaned to a level of $C_{s\,out}$. During washing the water becomes contaminated

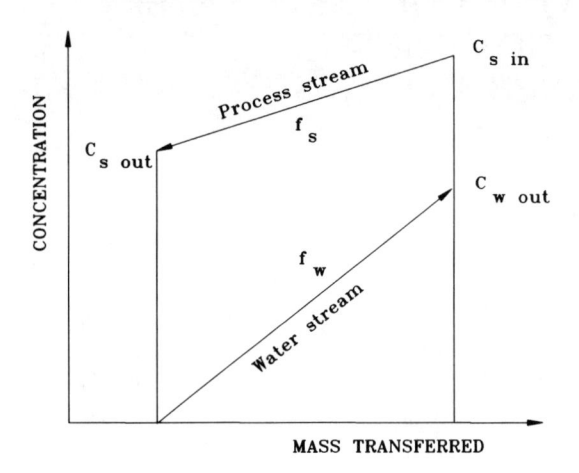

Figure 3.1 Graphical representation of mass transfer (washing) operation.

to a level of $C_{w\,out}$. This operation can be represented on a plot of concentration versus mass transferred (Figure 3.1). In this case the water flow rate is seen to be f_w. Now consider what happens if an effort is made to reduce the water being used for this washing operation. Figure 3.2 shows the situation where in order to reduce the use of water the only apparent degree of freedom is exploited — the outlet concentration of the water is allowed to rise. On reducing water use further and thereby increasing the outlet concentration, at some point a limit is hit — given by $C_{out\,max}$. This limit could exist for any number of reasons but typically may be due to contamination constraints, mass transfer driving force requirements or limits of materials of construction. Whatever the reason, there is some limit. To achieve this limit water use must be driven to the absolute minimum.

MULTIPLE WATER USERS

Now consider the same approach to multiple users of water. Table 3.1 shows a system comprising four separate users of water. The same approach is applied to each individual user in turn. Assuming that each user uses the minimum possible flow rate of water, the minimum water flows are calculated to be 40, 40, 32 and 6 t/h; this gives a total system requirement of 118 t/h of fresh water.

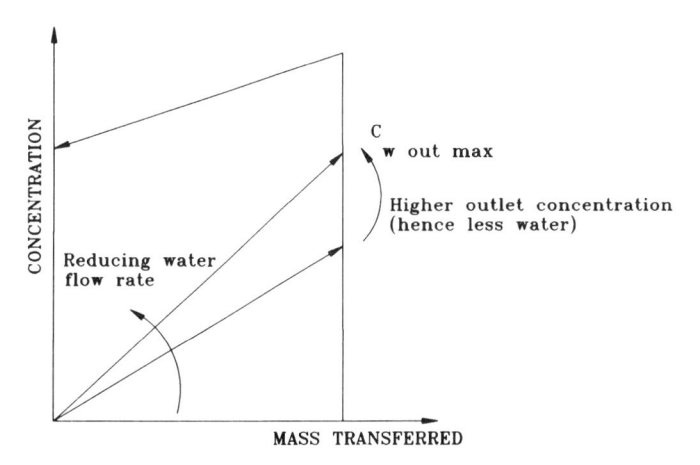

Figure 3.2 Minimum flow rate corresponding to maximum outlet concentration limits.

TABLE 3.1
Stream data for four water users

User	$C_{out\ max}$, ppm	Mass pick-up, kg/h	Minimum fresh water flow rate, t/h
1	400	16	40
2	400	16	40
3	1000	32	32
4	1000	6	6
		Total	118

Minimizing each user in turn appears to minimize the site's total freshwater requirements — or has it?

Before this question can be answered the problem needs to be reconsidered to see if any issues have been overlooked. Although a maximum allowable outlet water concentration has been specified, a maximum allowable inlet concentration has not been set. An initial reaction might be that such a

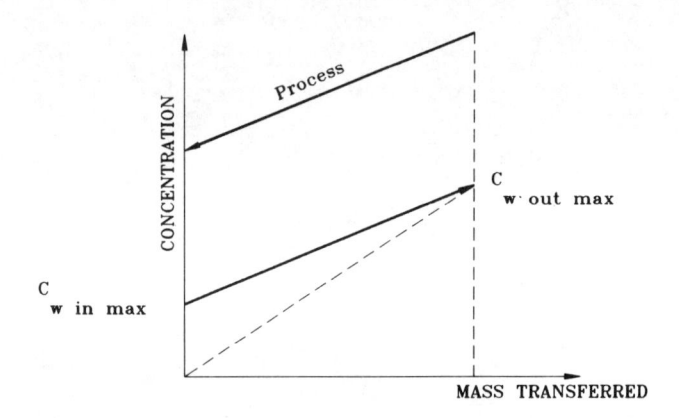

Figure 3.3 An alternative water profile suggests more water but accepts previously polluted water.

move gives no obvious benefit, in that utilizing slightly polluted water as the inlet supply gives less concentration difference (inlet to outlet) than when clean water is used. Hence, to remove the same amount of contaminant requires a higher water flow. Whilst this observation is true, the door is now open for possible water reuse (see Figure 3.3).

THE SYSTEMATIC APPROACH
Now consider dealing with the same problem, not as a series of four individual users, but as an overall system. To do this a construction is needed which represents how the system would behave if it was a single user and not four individual users — a composite curve. Firstly the data are completed by adding information on the maximum allowable inlet concentration for each user (see Table 3.2). A line joining the maximum inlet concentration to the maximum outlet concentration is defined as the limiting water profile and defines the boundary between feasibility and infeasibility — that is, any water supply eventually proposed which lies below the limiting profile will, in principle, be acceptable to the user, whilst any supply which exceeds the limiting concentrations will not.

Figure 3.4 shows the system composite superimposed on the individual limiting water profiles. The composite is produced by dividing the problem into concentration ranges — for example, range 1 would be from zero to 200 ppm, range 2 from 200 to 400 ppm, etc — and then simply adding together the pick-up of the individual streams in each concentration range. To then calculate the minimum fresh water requirement for the system simply requires the drawing

TABLE 3.2
Stream data with maximum inlet concentrations added

User	$C_{in\ max}$, ppm	$C_{out\ max}$, ppm	Mass pick-up, kg/h
1	0	400	16
2	200	400	16
3	200	1000	32
4	700	1000	6

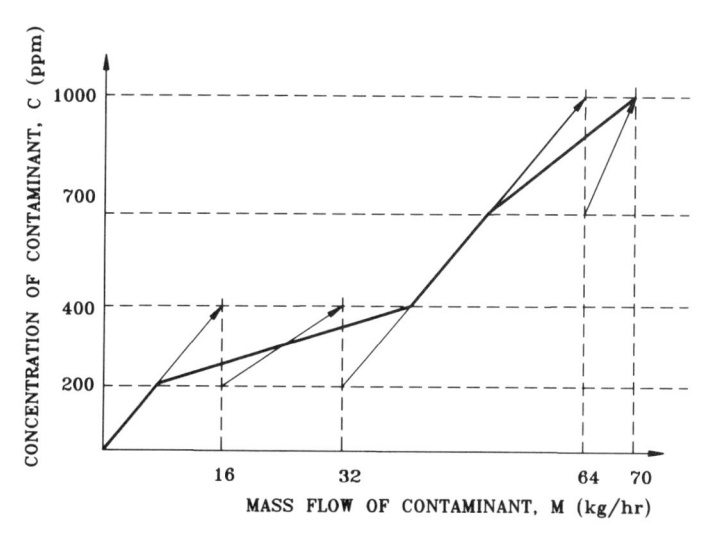

Figure 3.4 Construction of a composite curve.

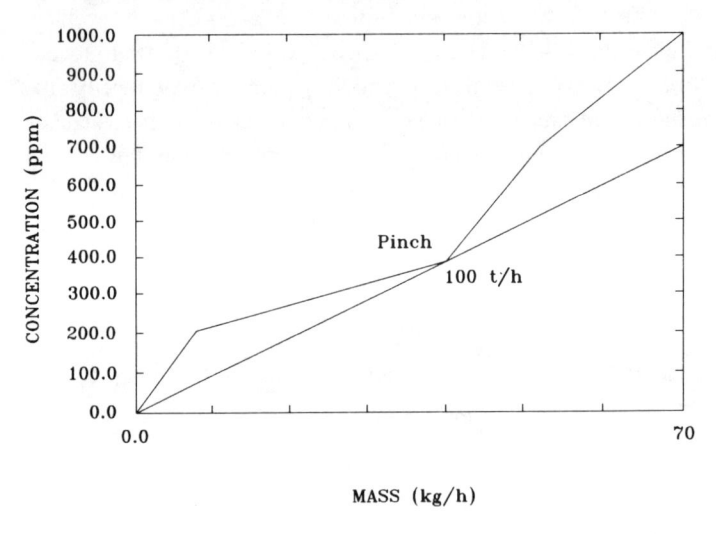

Figure 3.5 Minimum water supply is given by a line of greatest slope — a line which touches the composite at the 'pinch'.

of a water supply line of steepest slope (see Figure 3.5). The slope of this line is limited at one point — the 'pinch'. At this point one of the pre-specified limits is reached. In this case the minimum fresh water flow is seen to be 100 t/h. Techniques have also been developed which allow design for this target. One possible solution is shown in Figure 3.6. The design features reuse of water from process 1 into both process 3 and process 4. A bypass arrangement ensures that the water presented to process 3 is of sufficient purity.

Further water saving is also possible if a partial cleanup or regeneration is allowed. In Figure 3.7 the potential for further overall reduction in fresh water use by the appropriate use of a regeneration process is evaluated. A regeneration process is included which is specified as capable of removing 90% of the contaminant. It has been demonstrated[1] that the optimum overall performance of a regeneration process within the overall system depends on taking the water to pinch concentration before regeneration. Choosing a lower inlet concentration would leave scope for further water reduction. On the other hand, allowing the water to achieve a higher concentration will not reduce the water flow any further, but will simply increase the work (or cleanup) being carried out by the

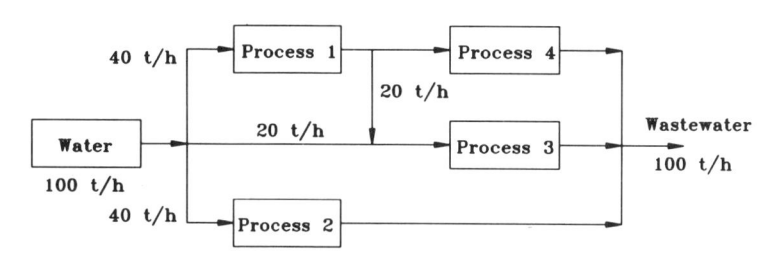

Figure 3.6 Design for water reuse.

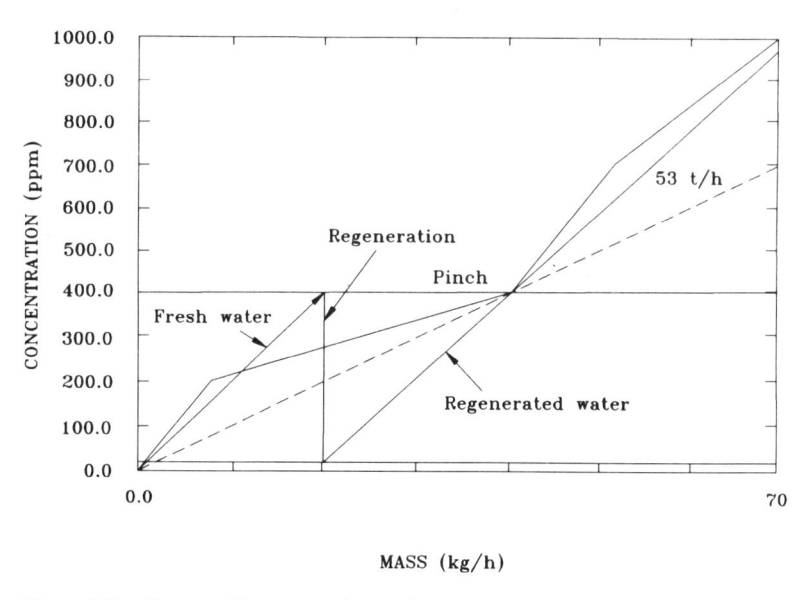

Figure 3.7 Target with regeneration and reuse.

regeneration process. In the example, incorporation of the appropriate regeneration facility allows the total fresh water requirement to be reduced to 53 t/h.

This approach can be taken further to look at the overall water saving benefits of alternative regeneration processes, recycling, etc. This builds up a picture or strategy of how water use can be reduced across any factory or site.

At each step, a target minimum requirement is calculated. This provides numerical information about the potential benefits of any modifications. If the potential benefits turn out to be large, a design can be produced which allows costing of the modifications.

An additional benefit of the composite curve approach is that it identifies the critical areas for future work. To make savings, those operations which are located at or near the pinch concentration must be addressed. Only in this region do the constraints cause any real problem in reducing water use. Once away from the pinch excess water is available. Hence, the technique has been very useful in identifying key areas where changes to the existing processes will give the largest benefits. At a time when resources are very limited it makes sense for industry to consider the benefits of a technique which identifies where to put our scarce resources for maximum benefit.

Although the methodology has been explained here on the basis of a single contaminant, the technique is easily extended to multiple contaminants, and all projects carried out so far have involved multiple contaminants.

THE RESULTS

The results to date of using this new technique have been impressive. Applications range from batch chemical plants through to oil refineries and petrochemical plants. Water savings potential has ranged from 30% to over 50%. Following a recent application of these techniques within Unilever the following statement was issued[2]:

'The final design solution, if implemented, will lead to a 50% reduction in fresh water demand and a 65% fall in wastewater production.'

REFERENCES IN CHAPTER 3
1. Wang, Y.P. and Smith, R., 1994, Wastewater minimization, *Chem Eng Sci*, 49 (7): 981–1006.
2. Hamilton, R. and Dowson, D., 1994, Pinch cleans up, *The Chemical Engineer*, 26 May, 42–44.

4. PROJECT CATALYST — A WASTE MINIMIZATION DEMONSTRATION PROJECT

Simon Clouston

Project Catalyst — a waste minimization demonstration project sponsored by the UK Department of Trade and Industry (DTI) and the BOC Foundation for the Environment and located in the North West of England — is believed to be the largest project of its kind ever undertaken anywhere in the world.

The project ran for 16 months, from January 1993 to April 1994, and involved 14 participating organizations from a range of industries. In total, the project identified potential savings of £8.9 million/year. Of these savings approximately £2.3 million/year were implemented during the project, with a further £3.7 million/year scheduled to be implemented within 12 months of the end of Project Catalyst — that is, by April 1995.

The majority of the savings identified were achieved at excellent financial rates of return. Zero cost opportunities — that is, no capital expenditure required — generated savings of almost £2.5 million/year and a further £3.0 million/year of savings were obtained for a payback period of less than 12 months. All the organizations taking part identified savings well in excess of the cost of participating in Project Catalyst, and all have continued the waste minimization 'journey' since the end of the project.

INTRODUCTION

There is ever-increasing interest in and concern about the impact of industrial and commercial activity upon the environment and hence there is increased pressure for action. This arises in various ways and from many different sources, but includes compliance with environmental legislation and greater awareness of environmental issues, both amongst staff and those external to an organization. For much of industry, the solution to waste and other environmental problems has traditionally been based on treatment and/or disposal. This approach is becoming increasingly undesirable, in part because it can involve significant capital investment, operating costs and the loss of valuable process materials. The costs of water, energy, effluent charges and off-site disposal have often been regarded as inevitable and/or insignificant. Many of these costs are set to rise significantly faster than the rate of inflation in the future. Also, tighter environmental controls will make 'end of pipe' solutions ever more expensive.

In this context the potential for waste minimization to help solve many of industry's environmental problems is considerable, but to date has been largely ignored.

PROJECT SCOPE AND OBJECTIVES
The overall objective of the project was 'to achieve the minimization of all forms of waste released to all the environmental media from a broad cross-section of UK business'. The 14 participants certainly represented a broad cross-section of business as illustrated by the following list:
- Borden Decorative Products Ltd — manufacture of wallcoverings;
- CMP Batteries Ltd — manufacture of lead acid electrical batteries;
- Colgate Palmolive Ltd — manufacture of body care and oral care products;
- J Crosfield and Sons Ltd — manufacture of speciality inorganic materials;
- Dunlop Ltd GRG Division — compounding of rubber and the manufacture of rubber-based products;
- H J Heinz Co Ltd — processing and packaging of food;
- Design to Distribution (D2D) Ltd (part of ICL plc) — computer and communications equipment assembly;
- J W Lees & Co (Brewers) Ltd — brewers and operators of licensed premises;
- Lever Brothers Ltd — manufacture of a range of household products;
- Manchester Airport plc — airport operations;
- Milliken Industrials Ltd — manufacture of carpet tiles;
- Pilkington's Tiles Ltd — ceramic tile manufacture;
- Royal Mail Collection — processing and distribution of mail;
- Stoves Ltd — manufacture of gas and electric cooking appliances and showers.

The participants received technical guidance and project management input from the project consultants who were WS Atkins–North West, March Consulting Group and Aspects International.

As well as including a very broad range of participants, a number of other features were considered as key aspects of the project. The project was:
- to look at the forms of waste released to all environmental media;
- to demonstrate the benefits of a systematic approach to waste minimization;
- to produce results which could be disseminated as examples of 'best practice' for UK industry to follow;

Manchester Airport plc — one of the 14 participants in Project Catalyst.

- to demonstrate the benefits of a 'club' approach to waste minimization with participants who would not normally have the opportunity of communicating on these issues.

METHODOLOGY

BACKGROUND
Although relatively new to the UK, a systematic approach to waste minimization has been applied successfully to many, many organizations overseas, particularly in the USA[1,2]. The methodology originally developed by the US Environmental Protection Agency has influenced much of the work undertaken subsequently, including the IChemE's *Waste Minimization Guide*[3].

The methodology adopted for Project Catalyst was in accordance with those identified above. It also took account of the lessons learnt in previous demonstration projects in Europe — for example, the Dutch PRIMSA project, the Landskrona Project in Sweden and the Aire Calder Project in the UK[4]. The project participants varied considerably in terms of their size, complexity and the nature of their business, and so the basic methodology was adapted to suit the site-specific requirements.

PLANNING AND ORGANIZATION

The very first task, in fact a precondition of participating in Project Catalyst, was to gain the commitment of the senior management within each organization. A project 'champion' was then appointed by each participant to lead and coordinate the waste minimization programme for its organization. The project 'champions' had to be sufficiently senior to be able to obtain the internal resources necessary to progress the work.

Site steering committees and task forces were then established to determine which areas of waste should be tackled and how. The wastes to be minimized varied from participant to participant but were generally selected on the basis of the following criteria:
- value of waste streams or types;
- perceived potential for improvement;
- impact on compliance with regulatory limits;
- impact on meeting internal management targets and objectives;
- potential for rapid successes to help promote belief in and commitment to the project;
- reduction of environmental impact.

It was important at this stage to try to identify all wastes, not just those which would be studied initially. This gave a 'hit list' of further waste issues to be tackled at a later date. As well as wastewater discharges, solid waste disposal, inefficient energy use and air emissions, other less obvious forms of waste were identified — for example, raw material losses, rework costs, poor plant utilization and unnecessary use of packaging.

AUDIT

Once the wastes to be minimized had been selected, an audit was carried out to identify and quantify the sources of loss. A 'mass balance' approach proved very effective for many, although certainly not all, of the wastes studied. Many of the participants were sceptical initially that sufficient data would be available to complete a balance. But in nearly all cases it was possible to produce an overall balance on a waste type to within $\pm$ 10%. This level of accuracy is normally adequate for prioritizing individual waste streams for further investigation.

OPTION GENERATION

Where a waste audit was undertaken, the loss data generated for each source was used to rank and hence prioritize the individual waste sources which would be selected for further investigation. Brainstorming sessions were used to generate preliminary options for improvement; typically between six and twelve ideas for each waste source were identified.

Each idea or option was then rapidly assessed for its likely technical feasibility and impact upon the waste stream being considered. This rapid ranking led to a prioritized list of minimization opportunities to be pursued. The opportunities were usually a mixture of those which could be implemented directly and those which required a more detailed feasibility study and/or design work before implementation.

The approach of waste audit followed by brainstorming was not the only way of generating options. In fact, Project Catalyst demonstrated that there are many instances when a different approach is more appropriate. Monitoring and targeting proved to be particularly useful in those situations where the waste arises as a result of variations in process efficiency.

Monitoring and targeting initially involved the regular collection of process and/or utility data. The raw data was then collated and presented in a variety of ways that enabled performance trends and comparisons to be produced. These outputs helped to identify the operations that were affecting the pattern of consumption or generation on site. This management information helped to target those areas of the process or the waste streams which required improvement.

Another important source of ideas and options for improvement was the work force. Members of the operational staff were always involved in the production of audit data and brainstorming sessions, but this often only involved a relatively small proportion of the work force. One of the important messages from Project Catalyst was that maximizing the involvement of staff increased not only the waste minimization opportunities identified, but also awareness and ownership of the waste minimization project.

The techniques used varied from site to site but included:
- short-term suggestion schemes;
- using existing quality circle arrangements;
- competitive team events.

Once identified and prioritized, the waste minimization opportunities were progressed using the participating organization's standard procedures for the implementation of any technical or management system change. Where required, more detailed technical and economic feasibility studies were undertaken and the funding obtained for those options requiring capital investment.

MONITORING AND TARGETING

Monitoring and targeting has already been discussed briefly as a technique for generating waste minimization options.

The use of monitoring and targeting techniques is also essential to the long-term success of a waste minimization programme. Once data has been collated, options generated and improvements implemented, it is necessary to measure the actual impact of the changes. Monitoring and targeting provide a measure of the actual effectiveness of implemented improvements; they also help to identify where operational or process changes have had an adverse effect on waste generation that had not been recognized at the time of their implementation. The techniques provide a historical record of improvement, as well as alerting management to any deterioration in performance.

Monitoring and targeting systems take many forms and do not always need to involve significant expenditure on computer hardware and software. It is important that a monitoring and targeting system is constructed with overall management objectives in mind and not just allowed to develop ad hoc. The experiences of the participants in Project Catalyst reinforced the following 'ground rules'.

Initially, decide what data needs to be collected, from where, how, how often and by whom. For example, if data on the towns water supplies to a factory are to be collected, how many meters are required and which streams should they measure? Are meters to be read locally or remotely? Ensure meters are specified accordingly. If meters are to be read remotely, ensure computer/data logging systems have sufficient capacity to receive all planned inputs. If meters are to be read locally, ensure that they are accessible and that someone is given the clear responsibility and sufficient time to take the readings.

Next, the collation and manipulation of the basic data needs to be considered. What outputs are required in terms of tables, comparisons, trend graphs, Pareto charts, etc? How will these outputs be generated and who will be given the responsibility, and resources, to ensure it is done?

Finally, how will the outputs be communicated and to whom within the organization? Also, how will the system outputs lead to action being taken if required?

By providing data on the success of implemented options, monitoring and targeting closes the loop and enables decisions to be taken on future priorities and targets for improvement. In the long term it is the use of a monitoring and targeting system that enables waste minimization to become a process of continual improvement rather than a 'one-off quick fix'.

CASE STUDIES

The following case studies have been selected from the 399 waste opportunities identified during Project Catalyst to illustrate some of the issues discussed in this chapter. The overall project results are summarized in a DTI report[5].

BORDEN DECORATIVE PRODUCTS LTD

The key waste issue targeted initially was the reduction of solvent-laden air emissions to atmosphere from the wallcoverings' printing lines. Solvents are used in the ink systems for printing the wallcoverings and a study undertaken prior to Project Catalyst had demonstrated that water-based inks could not meet product quality requirements at an acceptable cost.

The installation of an 'end of pipe' solution was therefore inevitable in order to comply with the conditions of the local authority Part B process authorization. However, Borden were keen to apply the techniques of waste minimization to reduce their emissions of both solvents and inerts to reduce the size and therefore the cost of any abatement equipment installed. A comprehensive programme of study and assessment work identified a number of ways of improving methods of ink handling and hood capture of solvent emissions. Careful attention to the detailed design of the extract ventilation systems for the printing lines enabled the total inerts flow to be significantly reduced.

As well as leading directly to some reductions in solvent emissions, it was estimated that the greatest financial benefit of the waste minimization programme will be a reduction of £1.5 million in the capital cost of the abatement equipment installed.

This example serves to demonstrate that undertaking a waste minimization programme does not preclude the use of other techniques or technologies to provide the optimum overall solution to a waste problem. Conversely, investing resources first in 'end of pipe' solutions invariably prevents the adoption of a waste minimization philosophy and so loses the significant financial savings which can be obtained by this approach.

CMP BATTERIES LTD

The major waste that was targeted initially by CMP Batteries was the liquid effluent discharges from the battery manufacturing process, which at the start of Project Catalyst totalled approximately 250,000 m^3/year. As well as trade effluent charges, this discharge represented a considerable cost in losses of purchased towns water and lead.

An audit was undertaken to identify and quantify all the significant sources of process liquid effluent and to develop a site mass balance for water usage. The quantification of the effluent sources provided a sound basis for prioritizing areas for further work. In the event it was discovered that there were about ten main sources which accounted for approximately 90% of the liquid effluent flow. Therefore it was feasible to generate and assess the waste minimization opportunities for all the major sources. This exercise was undertaken using a structured brainstorming procedure in order to:

- generate ideas for improvement;
- assess the likely impact of each idea on the waste source;
- assess the likely technical feasibility of each idea;
- rank the options identified rapidly to determine which should be implemented and/or assessed in more detail.

Having identified the favoured waste minimization opportunity for each significant effluent source, each potential project was reviewed for its contribution to the overall site strategy for water reduction. This ensured that the benefits of the waste minimization work were maximized and that improvements were not implemented in one area which would unwittingly prevent the implementation of improvements in another area.

The waste minimization opportunities identified included:

- replacement of a mixer exhaust wet filtration unit by a dry membrane filter to save 16,000 m^3/year of water and effluent;
- collection and settling and/or filtration of liquid effluent from two departments, enabling water and lead to be recycled to the process, to save 34,000 m^3/year of water and effluent and 75 tonnes/year of lead;
- reuse of treated site effluent for process cooling duties to save 64,000 m^3/year of water and effluent;
- connection of mixer cooler into an existing closed circuit chilled water system to save 12,000 m^3/year of water and effluent.

The implementation status of the opportunities identified for liquid effluent minimization is summarized in Table 4.1.

The crossflow filter at CMP Batteries Ltd, specially designed to separate oxide and water and subsequently recycle.

TABLE 4.1
CMP Batteries effluent minimization opportunities

Implementation status (at 1/4/94)	Number of opportunities	Saving Water, km^3/year	Lead, te/year	Money, £1000/year
Already implemented	5	75	30	74
Within 12 months	5	95	45	98
Feasibility study only	3	33	—	25

An important aspect of the implementation programme is that the final effluent quality is carefully monitored, to ensure that substantial reductions in effluent flow do not adversely affect the operation of the existing site effluent treatment plant.

Any such problems can often be avoided if minimization of the main pollutants in the effluent is addressed at the same time as reductions in the water flow. CMP Batteries addressed this issue by incorporating reductions in lead discharges to water into the programme of water and effluent minimization.

DESIGN TO DISTRIBUTION (D2D) LTD

Packaging waste was a major concern for D2D at its Ashton site, where it assembles, configures and tests computer and communications equipment. Packaging of components received by D2D from suppliers accounted for the majority of the 1000 tonnes/year of solid waste generated by the site.

With 5500 different component types being received from 1500 suppliers it was necessary to tackle the minimization of the packaging waste in a variety of ways. The solutions adopted to date include:

- over 50% of incoming component deliveries are now made directly to the assembly area where they are needed. The resulting reduction in handling has enabled the amount of packaging to be reduced;
- consulting and working closely with suppliers has led to an increasing number agreeing to take away their own packaging materials;
- extending the existing internal 'Kanban' system, which utilizes reusable transit cases, to some local suppliers has completely eliminated the requirement for packaging of these components.

The implementation of these opportunities had, by the end of Project Catalyst, already reduced solid waste levels by 100 tonnes/year and generated savings of £37,000/year in waste disposal costs, labour utilization and off-site storage charges.

D2D also devoted considerable efforts to developing the design and production methods of the packaging used for D2D's own products. One of the projects in this area was to design a system to enable 50% of product packaging to be developed.

The work undertaken during Project Catalyst identified potential savings of almost £400,000/year associated with reduction and recycling of product packaging.

J W LEES & CO (BREWERS) LTD

Due to ever-increasing water and effluent costs, water utilization at the Greengate Brewery was a major issue for the waste minimization work. J W Lees has been brewing at this site since 1828 and the numerous process extensions and modifications since then meant it was not possible to monitor water consumption for specific production activities. A comprehensive survey of the water distribution system was undertaken to identify appropriate locations for the installation of additional meters.

Once these meters were installed and the readings monitored for a period of time a number of problems became apparent in different areas, including:
- there was significant water usage when the plant was shut down;
- the pattern of water use did not correspond to production activity when in theory it should;
- water usage was consistently above that expected for certain plant locations.

As a result of the monitoring exercise, J W Lees was able to target areas for improvement. For example, it was discovered that solenoid valves on the emergency water supply to a refrigeration plant had failed. The repair of these valves eliminated a continuous source of towns water discharge direct to drain. In another area, the excessive use of towns water compared to the expected 'standard' usage was highlighted and reasons for it discussed with the operators. Over a period of several months, improved operational techniques reduced water usage in this area progressively to approximately 25% of its original value.

These improvements have already lead to savings of £30,000/year in water and effluent costs, demonstrating the contribution that monitoring and targeting techniques can make to the generation of waste minimization opportunities. As in this case, many of the opportunities identified by these means can be implemented without any capital expenditure, and lead to significant financial savings.

LEVER BROTHERS LTD

Lever Brothers had already undertaken a significant amount of waste minimization work prior to Project Catalyst. Audits had been undertaken to identify and quantify waste streams from the production facilities and many opportunities for improvement had been identified. Lever Brothers wanted to build upon the existing waste minimization initiatives to develop a continuing commitment to improvement.

A key issue was to raise the environmental awareness of all staff and to involve more personnel in identifying sources of waste and generating the options for improvement. This was implemented by the development of 'Spot the Waste' initiatives at the Warrington site. At first this involved some training of operating personnel to raise their awareness and understanding of what constituted waste and what the associated costs were.

The operating personnel from each production area — for example, a bottling line — were then formed into a number of teams. The teams were given a couple of weeks to generate as many viable ideas as they could to reduce waste in that area. For example, on the bottling line the types of waste included spilt product and damaged bottles, caps, labels, glue, utilities, cardboard boxes and other packaging.

At the end of the initiative a prize was awarded to each team that had generated more than 50 viable suggestions and an additional prize to the team which identified the single best waste minimization opportunity. For the bottling line, six teams generated a total of 609 viable ideas for improvement, although inevitably some of these were duplicated. An approximate initial estimate was that those most likely to be implemented would generate savings of at least £40,000/year.

This type of exercise is one of the best ways of generating improvement opportunities for the numerous smaller, often occasional, sources of waste — for example, the filling head that occasionally overfills bottles. Individually these sources may be small but can be so numerous that the overall cost of waste due to them is very significant. They are unlikely to be identified by the audit or monitoring and targeting approaches, unless the quantity of waste generated is large, because they are usually only known to the operators who work with the processes every day.

CONCLUSIONS
The opportunities for financial savings and reductions identified during Project Catalyst reinforced the message of previous waste minimization studies. Waste minimization:
- always saves money;
- always reduces an organization's impact on the environment;
- works for all types of waste;
- works for all types of organization — large or small, simple or complex, manufacturing or service;

- always generates opportunities for improvement, the majority of which will be 'no cost' or low cost — that is, payback in two years;
- always increases knowledge by enabling a familiar operation to be studied from a different viewpoint.

Project Catalyst also identified that a number of key factors are crucial to the success of a waste minimization programme:
- management commitment is essential — without it waste minimization will not succeed;
- organizations must understand the need for and the benefits of waste minimization;
- setting clear objectives is important to give direction to the initiative;
- sufficient, competent resources are required to enable work to progress;
- to achieve its full potential, waste minimization must be a process of continual improvement and not a 'one-off quick fix';
- site-specific management cultures and systems or technical issues do not prevent the successful implementation of a waste minimization programme.

REFERENCES IN CHAPTER 4
1. US Environmental Protection Agency, 1992, *Facility Pollution Prevention Guide*.
2. US Environmental Protection Agency, 1992, *Pollution Prevention Case Studies Compendium*.
3. Crittenden, B.D. and Kolaczkowski, S.T., 1992, *Waste Minimization Guide* (Institution of Chemical Engineers, Rugby, UK). A new edition of this book is due to be published later in 1994.
4. CEST, 1994, *Waste Minimization: A Route to Profit and Cleaner Production — An Interim Report on The Aire and Calder Project*.
5. Department of Trade and Industry, *Project Catalyst — Report to the Project Completion Event* (available from the Environmental Technology Best Practice Help Line — 0800 585 794).

5. ION-SELECTIVE MEMBRANES IN EFFLUENT TREATMENT AND PREVENTION
Jan Tholen

There are various ways in which ion-selective membranes can be used in the treatment of industrial effluents. This chapter gives an overview of a number of them.

In combination with electrochemical reactions or alone, these membranes can be used for the recovery of chemicals and metals, as is shown here in some practical examples drawn from modern processing industries.

WHAT IS AN ION-SELECTIVE MEMBRANE?
Ion-selective membranes have been on the market for over 30 years. They are based on polymers of materials such as styrene or polyethylene incorporating fixed and mobile charged groups. In, for example, cation-exchange membranes, negatively charged groups are fixed and positively charged groups mobile. Then, under the influence of a driving force (electric field or concentration gradient), positive and negative ions are attracted to the membrane, and the membrane is permeable for positive ions and rejects negative ions (Figure 5.1).

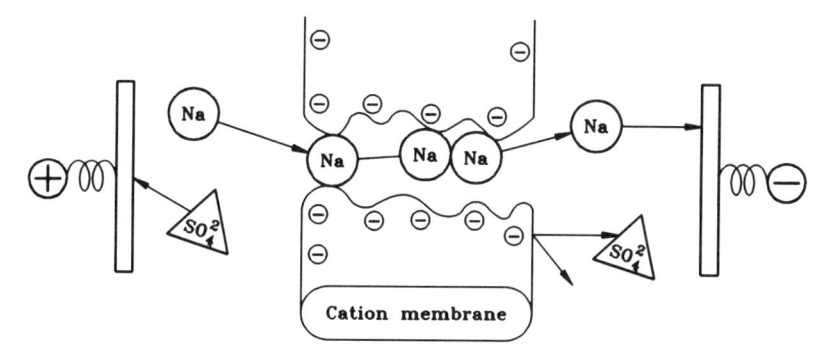

Figure 5.1   Cation-membrane.

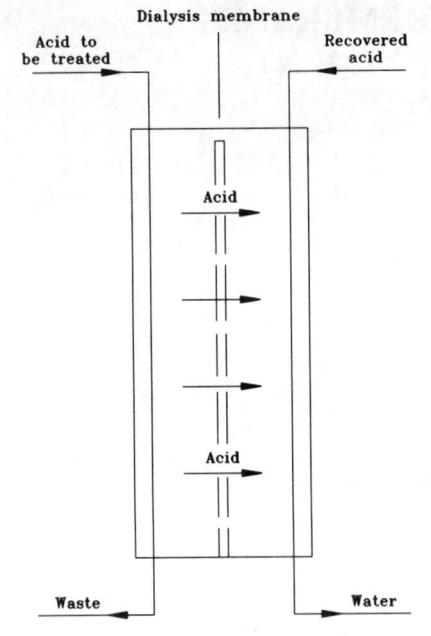

Figure 5.2   Dialysis.

PROCESS APPLICATIONS

The driving force for separation processes can be either a concentration gradient or an electric field. A number of different process names are used (depending on the driving force and the system layout); examples are dialysis, electrodialysis, membrane electrolysis and electrohydrolysis.

The following sections discuss these processes and give some applications. Consideration is also given to the economics of these applications.

DIALYSIS

In dialysis a concentration gradient over the membrane is used as the driving force. If, for example, an anion-selective membrane is placed between an acid-salt mixture and pure water, only the anions (acid-ions) will pass the membrane (Figure 5.2). But due to their high mobility protons will also pass the membrane. Using this method, acid can be separated from acid-salt mixtures. The mass transfer of acid is determined by the concentration gradient of the acid, the dialysis coefficient and the membrane surface. The dialysis coefficient depends

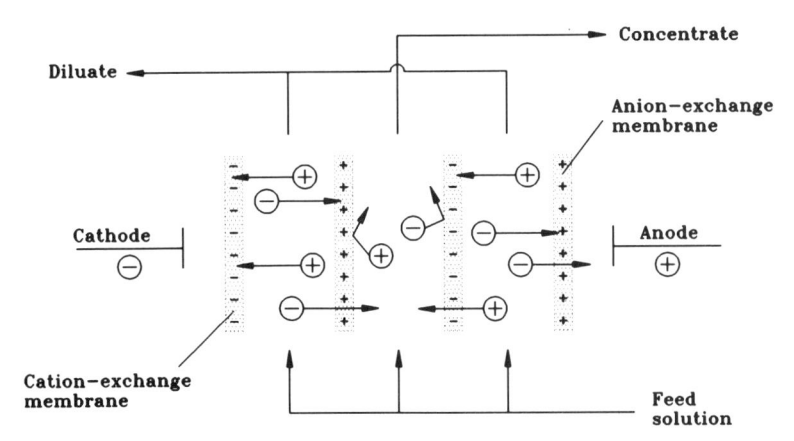

Figure 5.3 Electrodialysis stack.

on the type of membrane and the type of acid. Limited selectivity of the membrane leads to a minor leakage of the salt into the produced acid.

Applications of dialysis in effluent treatment can be found in the aluminum industry where etching baths can be recovered by the separation of sulphuric acid from acid-aluminum sulphate mixtures. Also the recovery of caustic soda or potassium hydroxide has become state-of-the-art technology with the recent introduction of cation-selective dialysis membranes.

ELECTRODIALYSIS
In the electrodialysis process the transport through the membrane is promoted by an electric field across the membrane. Diffusion is negligible by comparison and ion transport is directly related to the electric current. According to Faraday's law an electric current of 26.8 Ah is needed to transport 1 gram equivalent of a salt. In practice the electric current is higher due to inefficiencies caused by electrical resistance. Generally in electrodialysis a package of membranes is used with alternate cation and anion membranes arranged in a so-called 'membrane stack' (Figure 5.3). Anions and cations pass the membranes so that the original feed is desalinated and on the other side of the membranes a concentrate forms.

Typical examples of the application of electrodialysis in effluent treatment are the removal of salts from fluids, such as water, glycol and other process streams. They are all typified by the separation of dissociated from non-dissociated solutes.

MEMBRANE ELECTROLYSIS

Salt splitting

A well-known example of a membrane electrolysis process is the chlor-alkali process, in which chlorine and caustic soda are formed from a sodium chloride solution.

Treatment of waste from this process has been developed to produce pure base and acid from waste salt solutions. In industry salts are very often produced from the neutralization of an acidic or an alkaline fluid.

If the salt can be split into the original base and acid, full reuse is possible and waste streams are prevented. This process is attracting more interest due to increasing restrictions on disposal of salts such as sodium chloride and sulphate. Several membrane arrangements are possible, but the majority of applications need a so-called 'three-compartment cell' (Figure 5.4).

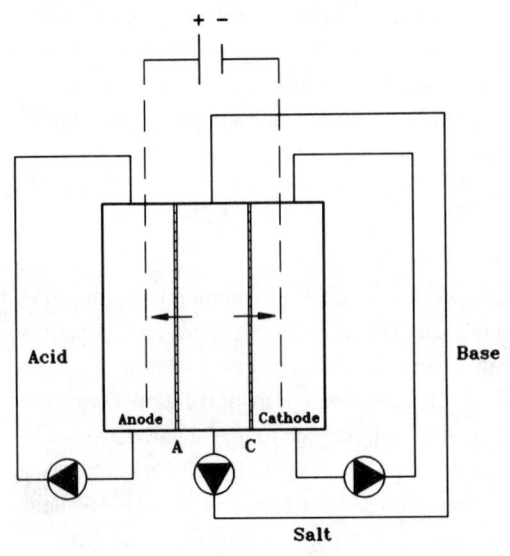

Figure 5.4   Three-compartment cell.
A = anion-selective membrane, C = cation-selective membrane.

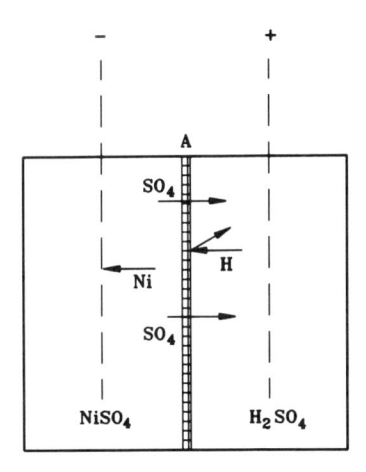

Figure 5.5   Metal recovery.

The salt solution is fed to the middle compartment and in the adjacent cathode- and anode-compartments pure acid and base are produced. Due to the electric field the anion is attracted by the anode and the cation by the cathode. Together with the H^+ and OH^- ions, formed by water splitting at the electrodes, acid and base are formed. The selective membranes prevent the transport of H^+ and OH^- ions. With special arrangements it is also possible to produce acids (weak organic acids or hydrochloric acid) which would normally be oxidized at the anode.

Metal recovery
If the salt is a metal salt it is possible to let the metal 'plate out' on the cathode. In galvanic processes generally this phenomena is well known without a membrane. But a membrane (Figure 5.5) often makes the recovery of metals more efficient. The formation of chlorine gas can be prevented. Iron is also often present in these metal salt solutions. Without a membrane an 'iron loop', the continuous process of reduction and oxidation of the iron at the cathode and anode (Fe^{2+}-Fe^{3+}) would minimize the efficiency, but a membrane blocks the loop. The conventional metal removal processes such as sedimentation and precipitation produce huge amounts of sludge and this creates a new waste problem. With membrane electrolysis these sludges can be converted to the pure metal or, if used in-line with the galvanic process, formation of sludges can be prevented.

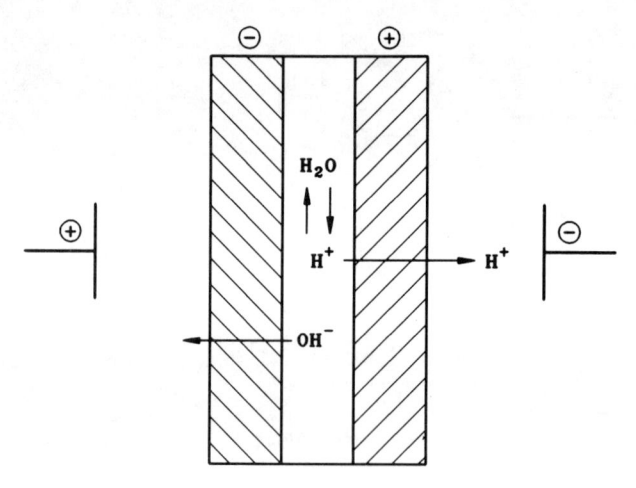

Figure 5.6   Bipolar membrane.

Membrane electrolysis is already used, both separately and in-line, on a commercial scale for applications such as the recovery of nickel, copper, lead and zinc. In chromium production this process is used in such a way that chromic acid is produced from chromates or Cr^{3+}-solutions.

BIPOLAR MEMBRANES
A bipolar membrane is a combination of an anion-selective and a cation-selective membrane with a very thin water layer between them. The water molecules in the membrane split in an electric field (Figure 5.6) creating a bipolar membrane that can take over the function of the cathode and the anode in the membrane electrolysis process as a producer of H^+ and OH^- ions. In an arrangement such as that in Figure 5.7 a salt can be split into base and acid. These membranes can also be stacked just as in normal electrodialysis.

The advantages of using bipolar membranes are:
- the possibility to use only one pair of electrodes for a package of membranes;
- the prevention of oxidation of the produced acid.

A major disadvantage is the limited selectivity of the membrane. In the produced acid and base, salt impurities at a level of g/l are always present.

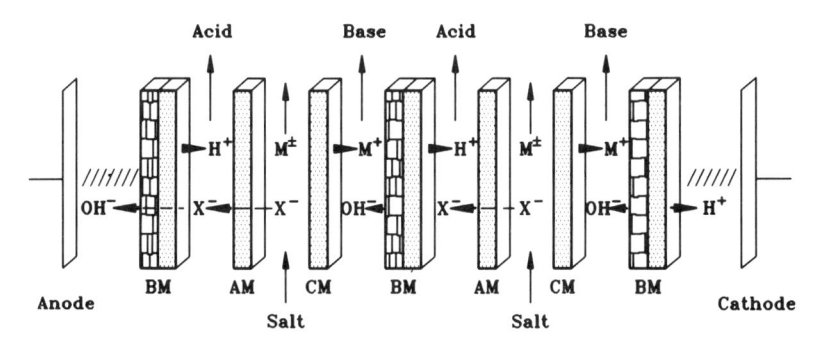

Figure 5.7 Bipolar membrane stack.

ECONOMICS OF THE USE OF ION-SELECTIVE MEMBRANES
This section presents some examples of the use of ion-selective membranes. Like most membrane systems, they are generally built as modular systems, which means that scale-factors are close to 1 and system capacities have only a minor influence on the economics.

All costs are given in pounds sterling.

SULPHURIC ACID RECOVERY WITH DIALYSIS
A mixture of sulphuric acid and aluminum sulphate can be separated in a dialysis cell (see Figure 5.8 overleaf). A 1000 l etching bath normally needs replacing every three months. The bath is treated with a dialysis unit with a capacity of 0.5 l/h, recovering 80–90% of the free acid and leaving a waste containing all the aluminum sulphate and only 10–20% of the acid. The investment costs of such a unit are about £3000, and operational costs — restricted to pump maintenance and membrane replacement — are £500/year.

Fluids which can be treated in this way include sulphuric acid, nitric acid, hydrofluoric acid, hydrochloric acid and chromic acid, caustic soda and potassium hydroxide.

GLYCOL DESALTING WITH ELECTRODIALYSIS
Table 5.1 overleaf shows figures for the desalination of a stream of 5 m^3/h glycol/water from 12,000 ppm salts to 700 ppm salts. The cost is £0.50 /m^3.

Figure 5.8   1 l/h dialysis unit.

Figure 5.9   Nickel recovery cell.

TABLE 5.1

Costs per year	
Membrane replacement	5000
Electricity	20,000
Others (including capital)	10,000
Total	35,000

NICKEL RECOVERY WITH MEMBRANE ELECTROLYSIS

From a galvanic process a waste stream with about 5 g/l Ni is treated in a membrane electrolysis cell. The rinsing waters are cleaned with ion exchangers and the regenerant, which contains nickel, is also treated in the membrane electrolysis cell (Figures 5.9 and 5.10). In a membrane electrolysis cell all the nickel is recovered as pure metal.

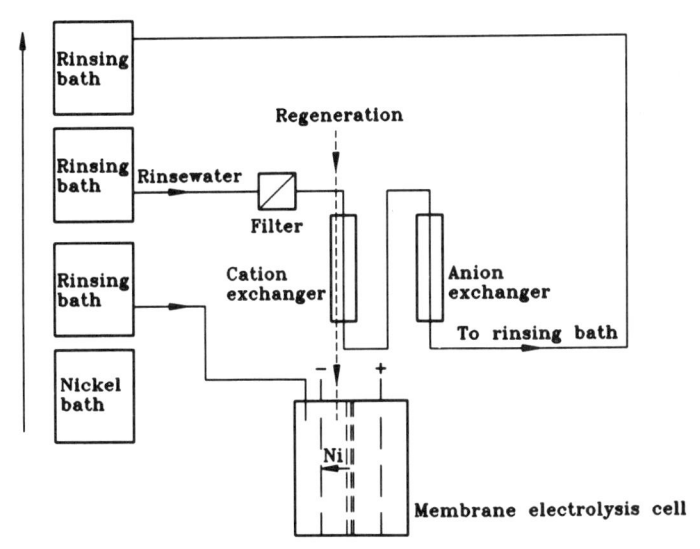

Figure 5.10 Nickel recovery system.

Costs (Table 5.2) include that of preconcentration by ion exchange. From the process 1000 kg pure nickel (99%) is recovered annually with a value of £3500. Prevention of sludge disposal saves £6000 per year, so that total investment costs are returned in one to two years.

RECOVERY OF ACID AND BASE FROM A SODIUM SULPHATE
WASTE STREAM
A waste stream with a 5000 ppm sulphate is neutralized first with caustic soda and concentrated in a reverse osmosis (RO) unit (Figure 5.11 overleaf). The con-

TABLE 5.2

Costs per year	
Membrane replacement	300
Electricity	300
Chemicals	1000
Others (including capital)	3000
Total	4600

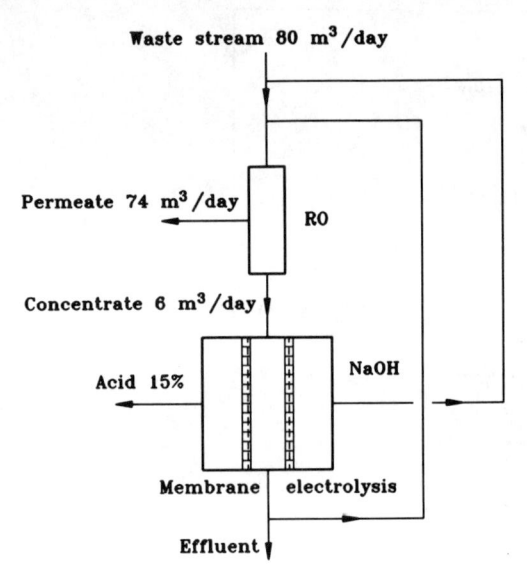

Figure 5.11 Sulphate removal system.

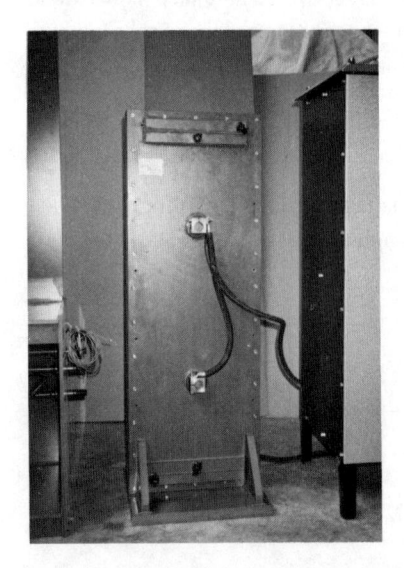

Figure 5.12 Membrane electrolysis cell.

TABLE 5.3

Costs per year including RO	
Membranes	20,000
Electricity	16,000
Others (including capital)	30,000
Total	66,000

Savings	
650 m^3 acid 15%	20,000
20.000 m^3 process water	20,000
Waste disposal	40,000
Total	80,000

centrate of the RO unit, with about 10% sodium sulphate, is then converted in a three-compartment electrolysis cell (Figure 5.12) to caustic soda and sulphuric acid, both as 15% solutions. The caustic soda is reused for the neutralization of the waste and the acid is reused in the process. The RO permeate is perfectly suitable for reuse as process water. Costs and savings are given in Table 5.3.

6. REVERSE OSMOSIS MEMBRANES IN COOLING TOWERS
Stuart Ord

The ROMIC (Reverse Osmosis Membranes in Cooling towers) project was initiated by ICI personnel working on membrane technology projects in 1992. It was favoured because of its wide potential application in the company, and so attracted funding from ICI Chemical & Polymers' strategic research fund.

Examples of this type of technology have been installed in parts of the USA and Australia, but not in Europe as far as we are aware. These developments have been driven by water scarcity and by disposal difficulties. It is believed that as waste disposal and raw water, especially potable quality water, become more expensive in Europe, then such technology will become increasingly attractive to Europeans. First hand experience of such a system is needed before detailed economic evaluations are possible.

BACKGROUND TO THE ROMIC PROJECT

Until 1993, ICI had a Membranes Section as part of its Chlor Chemicals Business Group. Whilst its primary aim was to develop and sell membranes hardware on the open market, it also had an interest in developing membrane applications in ICI C&P, particularly for reverse osmosis.

An investigation into the use of reverse osmosis (RO) to minimize blowdown from cooling towers was proposed to the managers of the strategic research fund in 1992. The External Collaborations Manager saw the potential to attract European Commission (EC) funding by developing this project as an APAS demonstration project with Separem.

Separem SpA is an Italian company based in Biella, in northern Italy. Its business is in both the manufacture of RO membranes and modules, and in the building of complete turnkey RO installations for customers.

From 1986–89 Separem was part of an EC-funded BRITE research and development consortium which developed an RO membrane with high resistance to dissolved chlorine (Applications of Membranes to the Textile Industry: Development of Specific Membranes and Processes: Project P1170, contract RI1B–73). The polyamide membranes thus developed are commercially available.

The EC was seeking follow-on projects to demonstrate the use and technology transfer of the products of BRITE projects. Therefore they supported the proposal for a project to apply these membranes to a cooling water which is treated with chlorine for microbiological control.

THE VALUE OF THE PROJECT
ICI hopes to gain a technology which will reduce its formidable bills for cooling tower makeup water, for effluent disposal, and possibly also for water treatment chemicals.

Presently, ICI C&P uses about 26 million m^3/yr of potable water, and about 166 million m^3/yr of lower grade waters such as river waters, in its processes. The water bill is about £17 million/yr. The proportion of this used by cooling systems varies from site to site, but a conservative estimate is that half of this water is used in open evaporative cooling systems.

Initial calculations suggested that a water reduction of at least 25% would be possible with this technology. Hence a saving of up to £2 million/yr was indicated, provided the technology could be made to work economically.

Separem hopes to gain an advantageous position in supplying membrane modules and systems to ICI and to other users of open evaporative cooling systems throughout the world. If ICI were to build RO systems at all its suitable cooling systems, then it is estimated that the value of replacement membranes would be about £100 k/yr. The market outside ICI is considerably larger.

COOLING SYSTEMS
Cooling water can be a very corrosive medium, and the operation of evaporative cooling systems must be such that corrosion of plant equipment by the cooling water is minimized (see Figure 6.1). This is done by the use of anti-corrosion chemicals. These chemicals suppress corrosion when operated within certain limits of pH, alkalinity and free chlorine.

As water is evaporated to reject the heat removed from the process to the atmosphere, the majority of contaminants in the water replenishing the system, both dissolved and suspended, become concentrated in the cooling water. Similarly, materials scrubbed from the air which contacts the cooling water in the tower further concentrate the cooling water. Unless these concentrations are controlled, the cooling water becomes highly contaminated and would deposit materials if the solubilities of various salts are exceeded. Control is normally

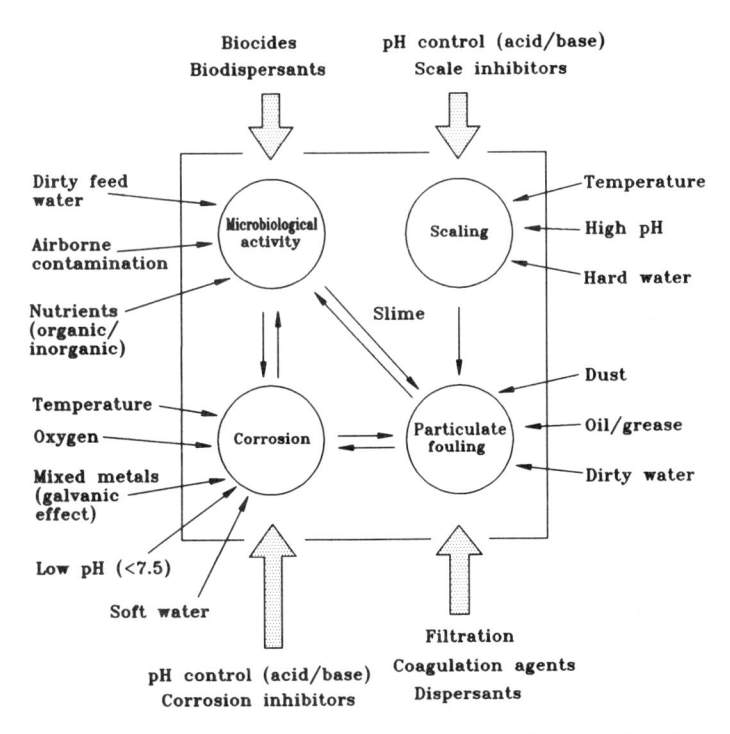

Figure 6.1 Factors affecting the performance and efficiency of cooling water and measures for their control.

achieved by purging water to drain, thus increasing the demand for makeup water (see Figure 6.2 overleaf). The amount purged is minimized, consistent with the overriding aim to control corrosivity.

Purged water therefore contains a relatively high level of dissolved and suspended solids, as well as dissolved water treatment chemicals. The ratio of the concentration of dissolved solids in the makeup water to their concentration in the purged water is known as the concentration factor. It is approximately equal to the ratio of makeup water flow rate to purge water flow rate.

Maximizing the concentration factor minimizes the makeup water consumption. It also minimizes the use of water treatment chemicals, but factors

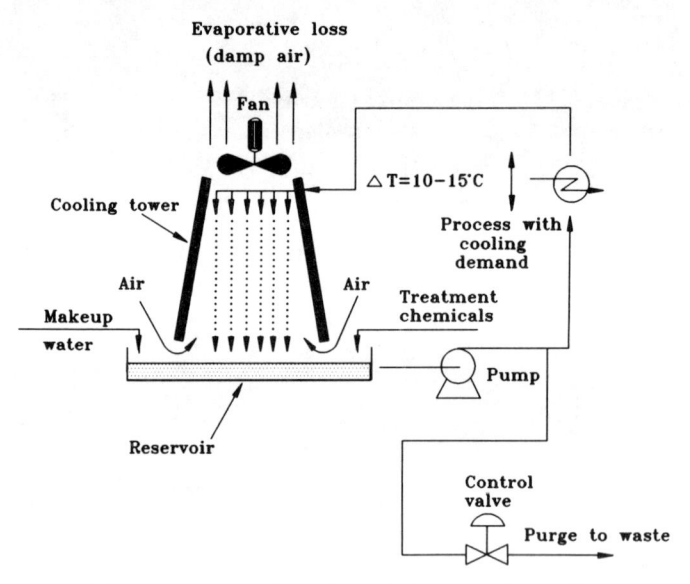

Figure 6.2   Schematic outline of a typical evaporative cooling system.

such as the half-life of the system have to be taken into account, as some chemicals lose their effectiveness after a certain residence time. Typically, concentration factors of the order of 2 to 5 are found. Lower values are used when lower grade water is used for the makeup — for example, river water — and higher values when potable water is used.

A few years ago, management paid relatively little attention to these figures, as the costs of the cooling system were relatively small. However in recent years, because of increasing prices and the need for all plants to seize every opportunity to reduce costs, concentration factors have generally risen and have been more tightly controlled.

Often the maximum concentration factor a plant can achieve is limited by involuntary water losses due to leaks. The steps which can be taken to improve cooling water costs are therefore:

(1) minimize involuntary water losses;
(2) increase the concentration factor to an optimum level by decreasing the purge rate;
(3) consider ROMIC technology.

A number of different water treatment systems are available; local conditions determine which is the best to use. In the past, many ICI systems were based on zinc chromate, but these have nearly all changed over to zinc phosphate based systems, which are more environmentally acceptable. Cooling systems in ICI are treated with proprietary chemicals, and the water treatment company (for example Nalco) manages the operation of these treatment systems.

REVERSE OSMOSIS

RO is a membrane process widely accepted for purifying and desalinating water. It is capable of removing dissolved ionic species down to very low levels. It works by applying the contaminated water to a thin sheet of a suitable plastic film through which water molecules are able to pass, but dissolved ionic species are retained. Thus the feed material is split into pure and impure streams (Figure 6.3). The process is not strongly ion-selective and, as a first approximation for the purposes of envisaging the process, the pure stream (the permeate) can be

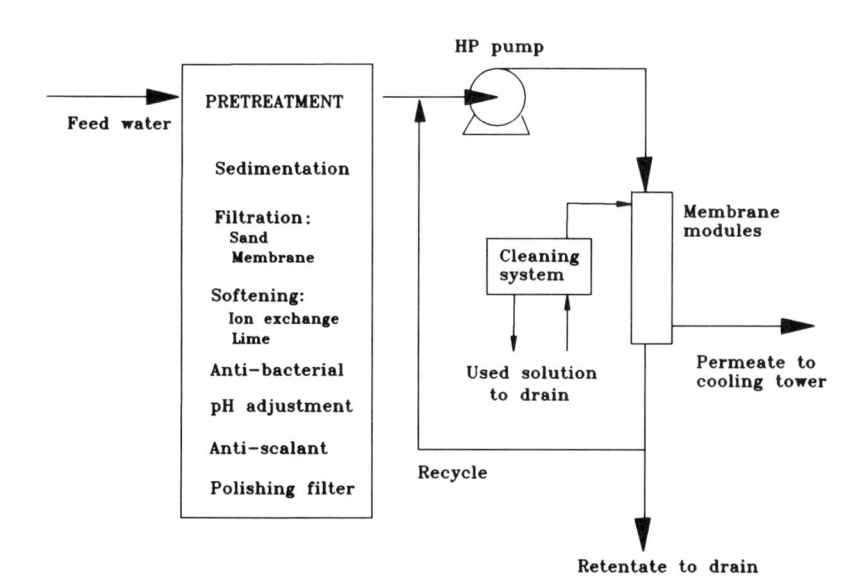

Figure 6.3   Reverse osmosis — general system.

considered to be pure water. In practice, about 95% of the dissolved material is retained in the impure stream, which is known as the retentate (or the concentrate).

The membrane can be thought of as a very fine polymeric filter. As would be expected, if the feed stream contains suspended materials, the membrane can become blocked, and productivity is lost. Similarly, if dissolved materials precipitate out onto the membrane as their concentration rises, or if microbiological growth occurs, then productivity is also lost.

Therefore the pretreatment of the water to a membrane system is vitally important, to maintain productivity and to obtain a good membrane life. The pretreatment section is often more complex than the RO section. It pretreats the water so that it is suitable for the membrane. If it fails to do so, the loss of productivity and lifetime of the membrane will have severe consequences to the project economics.

The pretreatment has to achieve conditions of temperature, pH, dissolved solids and particulate solids which optimize the overall system. In particular, the level of fine and colloidal solids must be controlled; an industry standard method called the 'silt density index' (SDI) is used to characterize this. The SDI test consists of passing the water through a fine-pore membrane filter under standard conditions and measuring the rate at which the flow through the membrane falls. Normally the method is used on waters with low solids contents, and SDI values fall in the range 0 to 6.67. A modified method had to be devised for cooling water as the raw water gave results well outside this range. The pretreatment system might also add chemicals to control the scaling potential of the water, or it could modify dissolved solids by ion exchange so that 'hard' ions are replaced by 'soft' ones.

Membrane systems are normally stopped periodically for chemical cleaning to allow deposits to be removed. During operation, the water velocity across the membrane faces is kept high to minimize fouling and to reduce the concentration profiles and boundary layer thicknesses. Therefore whilst a membrane system can be operated with up to 90% of the feed recovered as permeate, an internal circulation is arranged to make the recovery across the membranes themselves much lower, say 15–25%.

The pressure required for the process depends on the concentration of the retentate, as osmotic pressure has to be exceeded by a margin sufficient to get good productivity. The salt flux — that is, the passage of dissolved materials across the membrane — depends only on their concentration in the retentate. As a result, higher applied pressure, leading to a high water flux, will result in a higher permeate purity. However, this is at a significant cost of pumping the feed to the pressure needed, typically 20–40 bar g.

In very large systems, such as those used for sea water desalination, it is common to have a turbine/pump combination to recover energy from the waste retentate. Such an arrangement is unlikely to be economic on the scale considered here.

It is clear that there are a lot of factors to optimize in a membrane system. Experience shows that these cannot be done from theory, and extended trials are needed to understand the best combination of operating conditions and pretreatment technology to use to get the best overall system. This is especially true when operating on an environment about which little is known from these points of view.

REVERSE OSMOSIS IN COOLING SYSTEMS
It is possible to link RO into a cooling system in two main, distinct ways.

TREATMENT OF THE CIRCULATING WATER
Water is removed from the cooling circuit at a suitable place and fed to the RO plant (see Figure 6.4 overleaf). Generally it is better to take the water returning to the cooling tower, as it is warm. All RO membranes display improving production rate as the temperature increases, and the difference between using cooled water and return water can be as much as a 20% change in the membrane area needed.

This water is treated and the permeate is returned to the cooling tower where it takes the place of feed water, thus saving water. The concentrate is purged to drain. The amount fed to the RO system can be more or less than the purge rate previously used. However, the retentate flow will be less than the previous purge flow by the amount of water saving being made. Clearly the retentate flow rate should be minimized for maximum water savings, but the concentration of dissolved materials in the retentate rises in inverse proportion to this, and a practical minimum flow rate is reached just before the membrane starts to foul due to the deposition of dissolved salts.

All dissolved chemicals in the water being fed to the RO system are rejected to the retentate. This process variant rejects the water treatment chemicals in its feed to drain. The feed rate can be smaller than the purge rate from the cooling system before ROMIC, but only by 10–20%, so this arrangement does

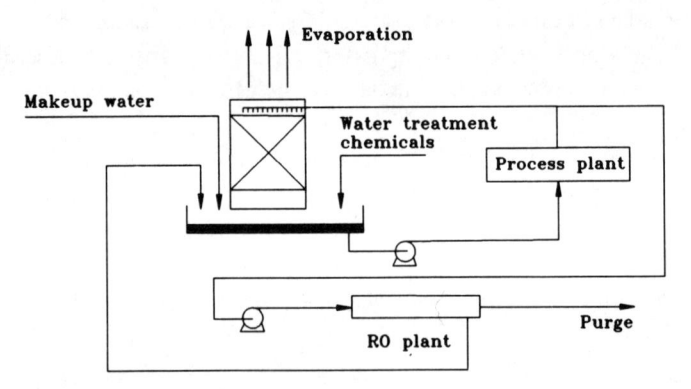

Figure 6.4   ROMIC — acting on purge.

Advantages over treating feed:
- feed rate reduced by concentration factor;
- better ultimate performance possible.

Disadvantages over treating feed:
- rejects treatment chemicals;
- more difficult pretreatment.

not reduce the environmental impact of the cooling system as much as the makeup water treatment arrangement. Presently this is not a very significant consideration, but it might become more so in the future.

TREATMENT OF THE MAKEUP WATER
In this process variant, the water feeding the cooling tower is treated before use, so that the cooling system is fed with a very pure water (Figure 6.5).

The cooling system purge is not treated, but is put to drain as previously. The flow rate of the purge can be much reduced from the original situation as the feed water treatment is removing most of the dissolved solids from the water being fed to the tower. A finite purge is still required to remove solids accumulated from the atmosphere, from plant corrosion, and to allow the minimum turnover of water treatment chemicals for them to be effective. As its

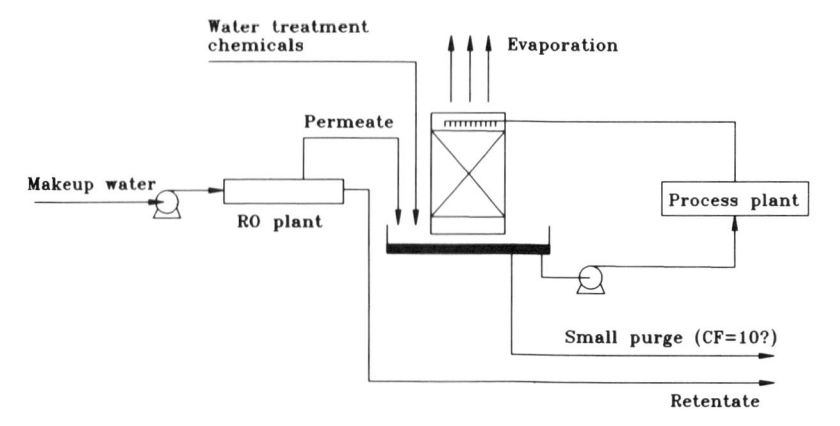

Figure 6.5 ROMIC — acting on feed.

Advantages over treating purge:
- minimizes loss of treatmen chemicals;
- simple RO pretreatment.

Disadvantages over treating purge:
- higher flow rate and membrane costs;
- process plant must have low water losses.

volume is much reduced, less water treatment chemicals are rejected to drain, and so the system can have a lower environmental impact than before, provided the RO system itself does not have damaging discharges, which is likely to be the case.

OUTLINE OF THE ROMIC PROJECT

The project is now coming to the end of its two-year life. It started in January 1993, and finishes in December 1994.

The phases in the work have been as follows:
- carry out laboratory work to screen water treatment chemicals against the proposed membrane materials, to ensure that a system is chosen which will not deteriorate the performance of the membrane material;

- modify an existing ICI RO rig and install it on a cooling system as a pilot unit;
- design, build and operate a new larger RO rig, to be installed on a suitable cooling system as a demonstration unit;
- optimize the mode of operation and operating parameters as far as possible within the timescale;
- analyse the results;
- report on the economics of the process, extrapolating to different locations where possible.

PILOT PLANT

It was decided to install a pilot plant unit on a cooling system at ICI's Castner-Kellner works at Runcorn. The unit was installed in a self-contained portable building. This allowed Research & Technology (R&T) Department personnel from Runcorn Heath to attend it regularly during the initial stages.

The equipment was an existing apparatus which had previously been used for investigating the use of RO to purify borehole water on another ICI C&P site. It was modified so that continuous unattended operation was possible. Two sand filters and two ion exchange beds, to provide base exchange softening, were arranged for duplex operation. The rig was designed so that either raw makeup water or cooling tower water could be processed, so that both methods of operation previously described could be studied.

Its capacity was designed to cover a large range of flow rates for characterization of the membranes, both under standard test conditions (which use low membrane recoveries without retentate recycle) and under operating conditions (where a high system recovery is needed). Similarly, alternative pretreatments were installed to allow testing of their relative effectiveness. The unit was also designed to allow any standard RO module to be tested to compare the Separem RO modules with others which are commercially available. The operating pressure was restricted to 20 bar g due to the limitation of the existing pump.

The unit was initially used to gather design information for the demonstration rig. Indicative productivities and rejections were found, and the decline of flux with time under different conditions was explored. Different pretreatment systems were tested, as were different cleaning materials. Some experiments were carried out with Optimem membranes (which are now sold by North West Water).

Later, after the demonstration unit had been commissioned, further tests were carried out on the pilot plant unit to guide experiments on the demonstration unit. It is presently being used to test arrangements which represent what we believe is the optimum economic operation for this variant of the technology, as this allows relatively stressful conditions to be studied without risking too many membrane modules.

PILOT PLANT RESULTS

Unfortunately there are a great many permutations of pretreatment, operating conditions and membrane management which can be tried. All sets of conditions need testing for significant periods of time before hard conclusions can be drawn, and so it is inevitable that some combinations cannot be tested. Ideally a statistical approach to experimental design could be used to plan the experiments in an optimal fashion. Significant pretreatment problems were encountered, however, which meant that the initial work concentrated on trying to achieve what was believed to be the minimum acceptable water quality for the RO part of the plant. Matters were complicated by the quality of the cooling tower water changing with time, especially when the Works carried out a cleaning of the cooling tower sump. This caused the SDI of the cooling tower water to increase dramatically for a period, followed by a reduction to below the original levels after the cleaning had been completed.

Whilst operation on potable feed water has not been a problem, the initial trials on cooling tower water indicated a problem with achieving the required SDI. Normally, pretreatment would be carried out to achieve an SDI of 5 units, and this was the initial target. However, the original sand filter bed, which had worked well on borehole water, was ineffective on the cooling tower water (see Figure 6.6 overleaf).

The sand used was a standard sand/anthracite mixture supplied by Memcor, who built the rig. The ion exchange resin used was type C100E. The water was filtered by 5 µm and then 1µm cartridge filters, which were intended to act as safeguards — for example, to collect fine ion exchange resin dust.

In practice, it was found that the sand filters had little effect. Some SDI reduction occurred across the ion exchange resin beds, and more across the cartridge filters. These filters, however, have a low solids capacity and needed frequent replacement; an improvement in the sand filter performance was essential. Analysis of the particle size distribution and discussion with sand filtration experts suggested that the solids being dealt with were at the limits of performance of the sand filters (see Figure 6.7 overleaf).

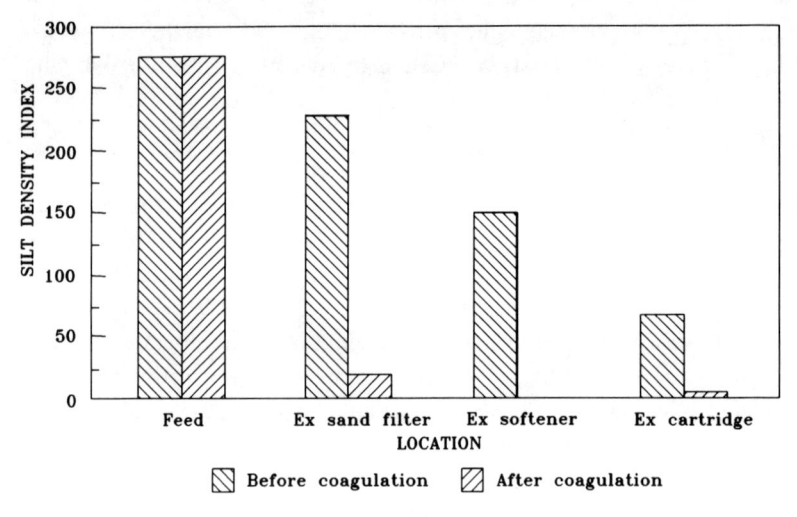

Figure 6.6   Silt density index — effect of coagulation.

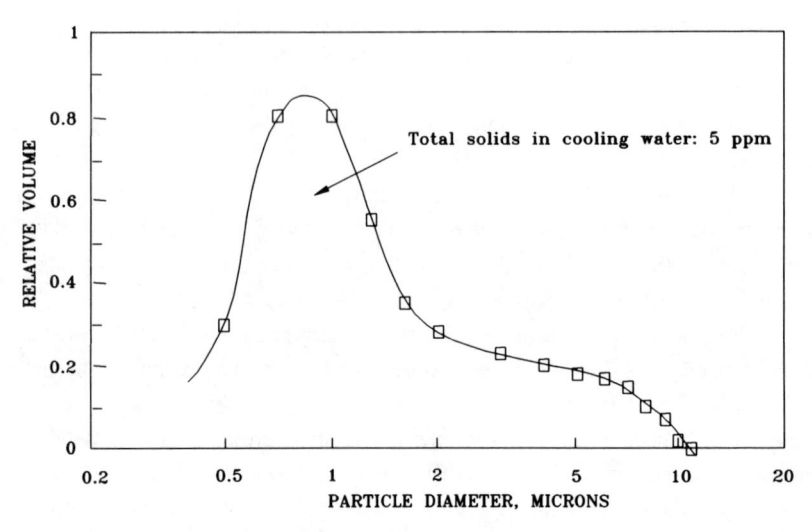

Figure 6.7   Size distribution of solids in cooling tower water.

IMPROVEMENTS TO THE SAND FILTERS

The sand filters were repacked with the finest grade of sand available, and a slight improvement in SDI was noticed. Layers of anthracite, 16–30 sand and 300–600 μm garnet sand were used. But these still proved inadequate, so the use of coagulants was investigated. Jar tests were initially used to select a suitable material. Use of this coagulant was successful in that the load on the cartridge filters dropped to a manageable level. However it was found that coagulant was passing through the sand filters after a few hours' operation, detected by a rising level of aluminium in the water going to the membranes. As aluminium ions can precipitate out and damage the membranes, frequent backwashing was needed to prevent their breakthrough.

Mass balances showed that suspended solids from the feed material only represented a small percentage of the total solids being removed during backwashing, the remainder being coagulant. Altering the coagulant level led to a decrease in performance of the filters. Therefore this arrangement showed a poor water economy. It had been anticipated that about 5–8% of the feed water would be lost to backwashing, whereas in practice it was significantly greater.

Experiments were then carried out to investigate the use of settling (sedimentation) as a pretreatment stage, either to replace or to supplement sand filtration. Jar tests were again carried out to select flocculants and coagulants. These were then tried on site. Nalco provided a settling tank for use with their settling agents. Whilst they were encouraging in the laboratory, they were ineffective in the field. Settling was therefore abandoned.

CROSSFLOW MICROFILTRATION

It was recognized that crossflow microfiltration (XMF) is a membrane technology which has the potential to solve the problem. It had been considered when the project was first planned, but the pilot rig already had sand filters. As time and costs were of the essence, and it was believed that sand filtration would be satisfactory, XMF was left as a fallback technology if needed. It was also believed that sand filters would be cheaper than XMF and would give better water economy. As the project progressed and the difficulty in reducing SDI arose, time and costs weighed against trying XMF. Eventually a practical way round the problem was found, and XMF was not tested. All the same, XMF may well become a viable pretreatment technology as its costs fall. Indeed, it is possible that a single stage membrane process, using an ultrafiltration membrane of suitable molecular weight cut-off, might be possible as an alternative to the 'traditional' two-stage process (that is, pretreatment followed by RO).

WHAT HAPPENED IN PRACTICE

In practice, several changes occurred to remedy the SDI problem. First, a routine cleaning of the cooling tower pond reduced the SDI considerably. Second, it was decided to see how long the membranes would last on water with higher SDI figures, and the results were quite encouraging. During this work, the SDI being fed to the membranes varied considerably from close to 5 to over 100. (The plant was not able to control this when coagulant was not used.)

This discussion primarily relates to purge water treatment. In the case of potable feedwater, the problem does not arise, and SDI figures of < 5 were obtained with sand filtration and cartridge filters.

Information was gathered on membrane productivities and their decline with time under differing conditions of pretreatment. Productivities were generally as expected, with potable water and low SDI purge water showing the best figures and lowest rate of decline. Varying feed conditions made it difficult to interpret some results. However it was found that:
- a peak flux of about 0.3 to 0.35 m^3/hr was obtained from the single 4040 membrane module at a pressure of 18 bar g, using 20% membrane recovery and 80% system recovery, on both purge and potable waters. This varied from membrane to membrane tried;
- flux decline rates were found to depend on a number of factors. These had different consequences:

Conditions	Relative flux decline rate
Softened potable water	1
Unsoftened potable water + Flocon 100 antiscalant	1.7
Low SDI, softened purge water	3
High SDI, softened purge water	6
High SDI, unsoftened purge water	Not yet tried; likely to be poor

- cleaning the membranes after a flux decline of up to 25% was successful, and the membranes were sometimes restored to better than original productivity after the first clean.

THE DEMONSTRATION PLANT
It was necessary to locate the demonstration plant at a tower whose evaporative load coincided with the size of RO system that the project could afford. It also had to be readily accessible to R&T staff and under good management control and phosphate treatment. Fortunately one such system, at Rocksavage Works at Runcorn, was available, and this was the one chosen.

The design was chosen to coincide with the overall features of the pilot plant. Sand filters and softeners, which were expected to perform better than those used in the pilot plant unit, were purchased from Italian vendors. The operating pressure was chosen to be up to 30 bar g, which Separem specifies for its modules.

Design, construction and commissioning of the rig took longer than anticipated. It is highly automated, being controlled and having data logged by a PLC/PC combination. The rig began beneficial operation in April 1994. The work was complicated by the delivery of membranes with higher productivity than the original ones used in the pilot rig, and the reduction in evaporative load on the tower to one below what the Works predicted when the plant was specified. An experimental programme to test as many parameters as possible is under way.

At the time of writing, few final results are available from the demonstration rig. However, much higher flux rates have been observed from the larger size (8040) RO modules — about 2 m^3/hr per module at 21 bar g and 20°C. The team will present its final report to the EC at the end of 1994. The summary form of the final report (Reference: APAS project 92–8, Contract no TPRO–0006) will be available from:
European Commission
Directorate General XII
Science, Research & Development
Directorate C: Industrial & Materials Technologies
Rue de la Loi
B–1049 Bruxelles
Belgium

LOOKING AT THE ECONOMICS

The economics of this technology depend on many factors. There are fundamental differences between treatment of the circulating water and treatment of the makeup water. In both cases there are further large differences according to the quality and nature of the water under consideration.

These differences primarily influence the technique required to pretreat the water prior to putting it across the membranes. River water clearly requires greater pretreatment than most potable waters, due to its higher levels of suspended solids and microbiological activity. Similarly, cooling tower water requires still greater pretreatment due to the higher concentration of dissolved solids in it, and the pickup of further non-soluble materials from the atmosphere. Given a certain input water quality, a variety of techniques exists to achieve a prescribed treated water quality. For example, suspended solids can be removed by sand filtration or sedimentation, according to their size and concentration. Similarly, sedimentation may or may not be a suitable preliminary stage, and the use of flocculants and coagulants might be investigated to assist with these methods. Crossflow ultrafiltration can also be used to remove solid particles, and a variety of physical arrangements is commercially available.

Finally, given all this, playoffs can be made between the level of pretreatment actually carried out — for example, the quantity of anti-scalant additives — against the membrane system performance, as characterized by membrane flux deterioration, membrane cleaning frequency, and membrane lifetime.

The economics are also strongly linked to geographical factors. Potable water prices vary greatly from region to region, and from state to state in the European Union. This ratio can be as high as 3:1.

Similarly, the cost of effluent disposal can vary greatly. In the UK, for large sites such as ICI's at Runcorn, the impact of putting ROMIC technology on even the largest cooling systems does not usually justify the claim that the technology moves an outfall from one size band to a lower one. This is mainly because all the outfalls are large, and the UK National Rivers Authority charging algorithm is roughly logarithmic on flow rate. Also, the cost of effluent disposal varies according to the location of the discharge, being three times higher for estuary waters than for ground waters in 1993–4. The technology will have much greater benefit to small sites disposing effluent to sewerage systems than to large ones discharging to coastal waters.

In summary, the technology has poorer economics:

- in regions using low grade water such as river water to feed their cooling systems;
- in regions of cheaper potable water;
- where water quality is poor, requiring extensive pretreatment;
- on large sites where savings from reduced effluent charges are small.

In the case of sites which are expanding or improving effluent standards, and therefore increasing the capacity of their internal effluent treatment facilities, a capital economy to offset the capital required for ROMIC technology can be claimed due to the reduced volume flow to these facilities.

CASE STUDIES

The case for ROMIC technology in various situations has been studied. A model has been written which allows data from the experimental work to be tested so that playoffs can be considered. At the time of writing, the operation of the demonstration plant had only been under way for four months, so much of the data has been taken from the smaller pilot plant.

The capital costs model used does not assume a cost proportional to flow rate, as is commonly assumed for membrane systems, but uses a factorial model with size exponents varying from 0.6 to 1.0. The capital costs have been minimized by assuming a very compact, modular plant layout.

The project has not been able to investigate alternative pretreatments such as lime softening or crossflow microfiltration. It is believed that as the latter becomes cheaper, it offers a possible improvement over coagulation and sand filtration assumed for the purge RO variant. The results of this analysis are shown in Table 6.1 overleaf. A number of cases are shown to reflect the results obtained from changing basic assumptions as well as changing the location of the RO system.

Calculations for sites using low grade waters such as river water have less attractive results, unless factors such as high water disposal costs or capital saving from reduction in size of future effluent disposal plants are included.

CONCLUSIONS

The use of RO to purify cooling water (that is, purge water but taken from the warm water returning to the cooling tower rather than from the tower pond) from a zinc chromate treated evaporative cooling system has been demonstrated to give useful water economies in line with predicted figures. Pretreatment has

TABLE 6.1
Results of economic analysis

Parameter	Case 1	Case 2	Case 3	Case 4	Case 5	Case 6
Water fed	Potable	Potable	Potable	Potable	Potable	Potable
Cost, £/m^3	Low: 0.6	Medium: 1	Medium: 1	High: 1.5	Medium: 1	Medium: 1
RO system location	Cooling tower water	Cooling tower water	Cooling tower water	Cooling tower water	Feed water	Cooling tower water
Cooling system evaporation, te/hr	High: 100	High: 100	Medium: 12	Medium: 12	High: 100	High: 100
Water saved, m^3/hr	25.6	25.6	3.1	3.1	14.3	25.6
Price escalation[1]	Yes	Yes	Yes	Yes	Yes	No
Project life, yrs	10	10	10	10	10	10
Tower CF with RO	n/a	n/a	n/a	n/a	15	n/a
Softening	Yes	Yes	Yes	Yes	No	Yes
Anti-scalant	No	No	No	No	Yes	No
Sand filters	Yes	Yes	Yes	Yes	No	Yes
Coagulation	Yes	Yes	Yes	Yes	No	Yes
Hypo dosing	Yes	Yes	Yes	Yes	Yes	Yes
pH trim	Yes	Yes	Yes	Yes	Yes	Yes
Pressure, bar g	30	30	30	30	30	30
Water, °C	30	30	30	30	15 (average)	30
System recovery	80	80	80	80	90	80
Saving £k/yr, year 1	105	194	21.7	35.1	138	194
Compound return on capital, % pa	24.9	31	22.4	27.5	15.4	26.2

Cost of disposal: £0.2 per m^3
Cost of water treatment chemical: £0.3 per m^3 purged from cooling system
Previous tower concentration factor: 4

[1] Escalation when applied: potable water prices rise 5% above inflation, effluent disposal prices rise 20% above inflation, for the period 1995 to 2000 AD

been found to be an issue in this case, but this can be dealt with by good houskeeping in the cooling tower and suitable pretreatment technology.

The use of RO to purify potable water so that water economies and reduction in the use of water treatment chamicals can be made has also been demonstrated. Pretreatment in this case is much easier than for cooling tower water.

Alternative pretreatment technologies exist which might show improvements on the economics reported here. Unfortunately the timescale of the project has not allowed these to be assessed practically.

Use of RO on cooling water gives considerably better rates of return than its use on feed water. This is mainly due to the fact that the size of the RO system is much greater for the latter than for the former. Whilst the cost of the RO system for feed water treatment can be minimized by the relative simplicity of the pretreatment needed, it is still about three times higher than the cost of the plant for the cooling water treatment. Part of this difference is due to the different water temperatures. The installation of a feed water preheater for the potable water is not economic.

The feed water case cannot achieve the water savings achieved by the cooling water case. This is because two purges are used — one the RO retentate, the other from the cooling tower. Dissolved solids levels prevent the higher recovery and concentration factor which would otherwise improve the water savings.

The actual returns obtained are shown in Table 6.1. The model assumes that water and effluent disposal costs will rise faster than the general rate of inflation. No case is shown for very low water costs — for example, use of river water for cooling tower makeup — as the low savings which the technology produces makes it uneconomic.

In regions of high water costs, very good rates of return are indicated. In regions where effluent disposal costs are higher than those assumed, or where water scarcity is important, the technology presents an even more attractive case. Further developments to reduce the capital and running costs might be possible before the end of this project.